LAND OF THE RISING SUN
(Numitna Thorakpham Japan Urubada)

LAND
OF
THE RISING SUN

(Numitna Thorakpham Japan Urubada)

SAMARENDRA CHONGTHAM

AKANSHA PUBLISHING HOUSE
NEW DELHI-110002 (INDIA)

AKANSHA PUBLISHING HOUSE
4649-B/21, Ansari Road
Darya Ganj, New Delhi - 110 002
Ph.: 23263193 / 9811582579 / 09435116718
Telefax: 011-23263193
Email: ektabooks@yahoo.com

Land of the Rising Sun (Numitna Thorakpham Japan Urubada)

First Published 2008
ISBN 978-81-8370-147-1

PRINTED IN INDIA

Published by Akansha Publishing House, New Delhi and Printed at Aryan Enterprises, Delhi.

To the

Nice People of Japan

CONTENTS

PREFACE

To sacrifice one's own life to protect one's honour has long been the tradition of Japan for thousands of years. The Japanese are a nation which considers it preferable to die then live in disgrace when he is defeated in a battle or a personal combat with a foe. It is a nation where harakiri is committed by disembowelling himself with a Samurai Sword rather than live as a defeated individual. Death in such a manner is considered as an act of bravery. Reminiscent of such suicidal traits of the Japanese could be observed during World War II. A few selected bands of pilot are there in the Japanese Air Force, undertake suicide-mission. They were known in the west as *Kamikaze*, which translates as 'divine wuid'. But in Japan they were known simply as the *'tokkotai'* or the 'special attack unit'. Strongly fortified enemy-position which could not be dislodged by aerial bombing, were destroyed by the *Kamikaze* fighter pilots by ramming their aircrafts to them, thereby sacrificing their own lives. During World War II, large American Warships which were immune to ordinary bombing from the air, the *Kamikaze* pilots, without caring for their own lives, undertook suicide-missions by directing their aircrafts into the funnel of the ship, thereby destroying the same.

In this way, the Japanese are a nation of proud, self-respecting people to whom the country's honour is uppermost in their mind. So much so – they prided that the sun shines first and foremost in their country. That is the reason why Japan is also known as the 'Land' of the 'Rising Sun'. The name of the flag of Japan is called Hinomaru — a round red mark in the centre against a white background. *Hinomaru* itself means 'round red disce of the sun'. A wonderful nation, indeed ! Living for three months from July 21-Oct 22, 1995, with these brave and hard-working people, I had the rare opportunity of observing from a close range, their life-style, customs and habits. And I record here my feelings and impressions of Japan, as I saw them with my own eyes.

First of all, I must apologize to my readers that I had repeated certain fact and points again and again. Not that I am unaware of the flaw in this book, yet I had decided not to delete it. My book is the factual records of my feelings and impressions then and there itself.

There is quite a difference between our trip to Japan and that of other groups from Manipur, who had visited Japan before us. There may even be persons from Manipur who had stayed in Japan for a year or so. But we never stayed in one fixed place. Travelling for three months in a bus, we had the rare opportunity to see not only big cities and ports but also small villages and islands in the midst of the ocean. We stayed with Japanese families and shared their meals. Wherever we went, be it a big town or a small village, I had the opportunity of speaking a few words to the assembled audience, regarding the land and people of Manipur. When I enumerated in detail about Manipur, there were many Japanese who were keen to visit Manipur. In fact, two groups of Japanese tourists, including two women, came here after we returned to Manipur shortly. In fine, our recent tour to Japan helped to strengthen the tie between Japan and Manipur.

While writing this travelogue, Ratan Thiyam gave me lots of encouragement for which I am grateful to him. Last but not the least, I shall be failing in my duties unless I record my gratitide to Sri Tayenjam Bijoy Singh, for going through the MSS of this book and making necessary corrections in it.

SAMARENDRA CHONGTHAM

LAND OF THE RISING SUN
(Numitna Thorakpham Japan Urubada)

An Observation

I have gone through the MS of the above title prepared by Samarendra Chongtham recounting in his own racy style what he has personally seen of the life, culture and material achievements of the great Japanese people as he had the rare occasion of traversing the length and breadth of the that charming country, Japan (covering in all 40 cities, towns and villages) as the manager of a Manipuri cultural troupe for full three months from 21 July to 22 October in 1995. Needless to say, Japan and her people are household words in Manipur since the Japanese army came down on the soil of Manipur to fight alongside the INA for India's independence during World War II. Now the present work has brought home the real Japan to Manipuri readers in their vernacular language thereby also throwing open a clear vista of the aspects of interesting similarities between the cultures of the two peoples – the Japanese and Manipuris. So, in a word, it is a bridge between the two remotely located peoples.

The attempt is laudable as it will be the first of its kind in Manipuri language and if brought out it will definitely find a place in Manipuri literature as a travelogue.

Prof. Ch. Manihar Singh
MANIPUR UNIVERSITY
DEPT. OF MANIPURI
CACHIPUR, IMPHAL - 795001
Phone :2220787
Extn : 252

THE WORLD PERFORMING ART FESTIVAL

3rd NIWABISAI, 1995

YAKUMO-MURA, JAPAN

It gives me great pleasure to record our appreciation and thanks to MR. CHONTHAM SAMARENDRA SINGH, who had participated in the World Performing Art Festival, the 3rd Niwabisai, 1995, at Yakumo-mura, Shimane Prefecture, Japan, on 16 September, 1995. I wish you success in life.

Yasuo Seko
Director,
Niwabi Festival Committee

Kumanotaisha,
Yakumo-mura,
Shimane Prefecture,
Japan
the 16th Sept., '95

POST INDIAN FESTIVAL COMMITTEE

CERTIFICATE

To Mr. CHONGTHAM SAMARENDRA SINGH

Your 40 performances in the cities and towns including Tokyo, Yokohama, Fukuoka and small island Aogashima(with 200 population), and Universities all over Japan are great success. Your contribution for cultural exchange between Japan and India is great. You introduced various Manipuri culture, like dance, cookery and so on.

This is the first opportunity for Japanese to close contact with Manipuri dance. The Post Indian Festival Committee will pay the respect for your great achievement.

Mithila Museum
Tokamachi, Niigata
Japan
October 18th, 1995

(TOKIO HASE...
Representati...
&
Director of
Mithila Museum.

Dear Indian Delegation of Universiade 95'

It was very nice to meet you in the frame of the cultural festival held at Hakata Public Health Office in Chiyo school range.
We thank of your visit may expand grass root friendship between India and Fukuoka-Japan.
Please write letters to us for more extension of friendship.

JAYA JAYA BHARATE - NIPPON friendship

Chairman of
Universiade 95' Support Society of Chiyo
Kensuke Sakimura

K. Sakimura

Address: 1chome 21-16, Chiyo ,HAKATA-ku ,FUKUOKA city ,JAPANA 812
Tel : 81-92-641-4404

1

THE JOURNEY BEGINS

The internationally renowned theatre company, Chorus Repertory Theatre of Manipur had assigned me to lead a group of Manipuri artistes for giving performances of Manipuri folk dances, *pung* and *dholok* (Manipuri percussion instruments) in Japan for three months.

We, a team of nineteen (of which 7 are female artistes and 12 are male artistes) including myself as Manager and Team Leader, had toured Japan and performed in 42 different cities. Exactly at mid-night of July 21, 1995, we left India by Singapore Airlines from Dum Dum Airport, Kolkatta. We had opted for this Airlines since the fare was comparatively cheaper and further offered concessional fares for group-bookings. As soon as we were airborne, dinner was served to the passengers. It was 1 AM when we started having our dinner inside the aircraft. The air hostesses were all very young girls, perhaps, aged about 22-23 years and they spoke good English. They looked so much like our Manipuri girls in their appearances. They greeted us with warm smiles, not the usual cosmetic smile of airhostesses, but with genuine warmth and feeling to one of their own kind. Instantly, they made friends with our girls.

Some of them even went ahead exchanging their names and addresses.

We reached Singapore in the early hours of July 22 by 4 AM I.S.T. (Indian Standard Time) but 7 AM by Singapore time. It was still pitch dark. We changed our aircraft at Singapore and by 11 AM left for Osaka, Japan. The food served to the passengers inside the aircraft was indeed, a wholesome fare and tended to forget the monotony of a long flight - fish, meat, sea weeds and all sorts of drinks in plenty. Just before lunch time the airhostess pushed her trolley filled with liquor of all sorts in the gangway. Passengers were free to choose their favourite brands - virtually a dream come true for drinkers - whisky, brandy, sherry, vodka, white rum, red wine etc., to mention only a few. You could drink to your heart's (Stomach's !) content; of course, free of cost. You need not pay anything extra. I helped myself to a little Vodka; it tasted like our own country brew, Sekmai. All of this were quite a new experience for us; it was the first International flight we had ever taken.

The flight from Singapore landed at Osaka Airport at 7 p.m. Two young men working in the Mithila Museum, Yoshihiro and Keiuske - received us at the Airport. They brought two vehicles driven by themselves - Yoshihiro with a 30 seater air-conditioned micro-bus (our own version of mini-bus) and Keiuske with a pick-up van for our luggages and costume boxes. We ferried our own luggages and the costume boxes and loaded them into the pick-up van. In Japan there are no porters; you have to do it yourself. By 11 PM we left Osaka Airport and headed for Mithila Museum. Our group came to Japan on the invitation of Tokio Hasegawa, Director of the said museum, located at **Oike** village, deep in the hills of Tokamachi city of Niigata Prefecture. All our dance performances in different locations of Japan numbering more than 40, were arranged by Tokio Hasegawa.

Mithila Museum, Oike, Tokamachi, Japan

During our drive from Osaka to Mithila Museum I enquired to Yoshihiro, the young man (aged about 24-25 years) who was driving the micro-bus about the distance up to our destination. Pat came the reply, "Not much, may be only about a thousand Kilometers". Driving non-stop throughout the night, both the driver and the passengers were all tired. We therefore, stopped at a wayside parking lot, prepared tea ourselves and took rest for a while. Our bus was loaded with enough provision of tea leaves, sugar, milk, biscuits, etc. Yoshihiro told me that he was feeling tired and needed some rest. We were even more keen that he should certainly take the much likely needed rest, fearing the pos-sible consequence of a sleepy driver at the wheel. All of us had a short nap inside the bus and then continued our journey. By then the sun appeared on the eastern horizon while our watches showed only 2 a.m. (I.S.T.). That was the first sun we ever set our eyes on in the Land of the Rising Sun; the day was July 23.

Having seated inside the bus for 12 long hours, we arrived at Mithila Museum by 11 a.m. The distance we had covered, as I guessed, should have been nothing less than 1200 Kms because our micro-bus ran at the rate of 120-150 Kms per hour. At Mithila we were greeted with a hot meal of rice, *dal* and *subzi* prepared by the staff of the Museum. As soon as we finished our meals, we made a hurried preparation for the next lap of our journey to a nearby island called Sadogashima by ship. We were baffled not the least since Hasegawa-san or none of his family members were seen around to welcome us on our arrival at Mithila Museum.

The mystery would soon be unfolded.

2

SADOGASHIMA

Hasegawa-san* was already there at the port at Sadogashima, along with his wife and two children waiting for our arrival. His 14 year-old daughter Raina and son Munna (8 years) welcomed our group with cheerful smiles. It took about 2 hours by ship up to Sadogashima. Hasegawa-san accompanied us. In fact, he was our constant companion and guide throughout our long and extended tour of Japan.

The two vehicles in which we came from Mithila Museum, the micro-bus and the pick-up van drove straight into the cargo hold of the ship. After reaching Sadogashima, we proceeded to a small fishing village called Akadomari. There our girls prepared dinner and retired early for the night.

During the nearly two hours ship voyage to Sadogashima, I had a long chat with Tokio Hasegawa-san. From him I learnt that the Manipur Cultural Troupe was sponsored, not by the Government of Japan, but by

---

* *It is not customary in Japan to use the English prefix Mr. and Mrs. before one's name. Instead, the word 'san' is invariably suffixed to names, both for males and females. It is equivalent of Sri and Srimati or Mr. and Mrs.*

the collective effort of artistes and art-lovers belonging to various cultural organisations. That was one reason why our accomodation in Japan could not be arranged in expensive hotels. We led a spartan life, staying in public halls, gymnasiums, clubs, Buddhist shrines etc.

It was in the village of Akadomari that we had our first experience of the Japanese way of life - staying in a Japanese home and eating there. Our greatest surprise there was the mode of using sandals or slippers inside their houses. Let us call it sandal-code. Nobody can enter a Japanese home or a public hall with their footwear on. It is their custom. Everyone has to remove their footwear in front of the door and change to slippers already kept there for the purpose. There are always about 10 pairs of slippers kept near the front door of every house.

Removing the shoes and other footwear before entering someone's house, however, is not a new custom for the Manipuris. We do follow the same custom and tradition but the Japanese stretched this a little too far. In Japan you have to leave your shoes and change into a pair of new slippers when you enter a house. As you enter with a new slipper on, another set of new slippers await you in front of the kitchen door. While going to toilet or lavatory, you have to leave the kitchen slipper and change to toilet slippers. This was the first culture shock we had in Japan! Thus while staying in a Japanese home, administering one's slippers is a tough task and for the first few days of our arrival in Japan we did practically nothing except slipper training! Since we were not familiar with such a quaint custom of changing slippers every alternate minute it was, no wonder, we often blundered and entered the toilet with the kitchen slippers on or *vice versa*. Our amused and watchful

Japanese hosts had a tough time correcting our errant artistes. Such were the funny incidents we experienced for the first few days. But as the days passed, we became familiar with their habits and customs and were quite at home.

A Japanese festival at Akadomari

We presented our first performance in Japan in this village at 7 p.m. of July 24. The show was arranged in the spacious Gymnasium of Akadomari village. No stage or even a dias was constructed inside the huge gymnasium which was almost as big as our Govindji Mandap. The audiences sit on its plank floor covered with mats, made of fine reeds, called Tatami in Japanese language. Tatami is part and parcel of every Japanese home. During daytime, they sit on it. And at night, they spread their bedding over it and sleep. Unlike the Indians, the Japanese do not use elaborate wooden charpoys (cots), mattresses, pillows and mosquito nets, etc. Theirs is very simple; they all sleep on the Tatami spread over their plank floors. The next morning, all the accessories like linen, pillow, mattresses, etc are rolled up and put inside the built-in cupboards and the rooms are utilised as drawing rooms during daytime. This is because, by and large, Japanese houses are small and there is always a problem of space. Their houses are fitted with sliding wooden doors. Land prices in Japan are exhorbitant and as such, only a few people can afford the luxury of a large and spacious bunglow, as we do in India. Secondly, Janpan is an earthquake prone country and as such the houses are made of light building materials, mostly of wood and timber. I scarcely came across any dwelling house made of brick and mortar and R.C.C. pillars etc, as we do in our country.

After our performance at Akadomari, the villagers and their energetic Mayor, Hideo Isezuki-san gave a reception for our group in which elaborate food tables were laid for us. But none of us could relish the rich fare. Their food might be rich in protein and calories, no doubt, but their cooking medium was different from ours. Their fish-fry, soya-sauce and various other items prepared mostly in soya oil were bland and tasteless to the Manipuri palates. Although they had laid the very best of Japanese cuisines, it was not suitable for our taste. Needless to say, we returned that night to our lodge completely famished and starved. So after midnight we cooked food Manipuri style at our lodge; there was no dearth of victuals as it was provided by Hasegawa-san. All the famished artistes joined in the cooking. Only after a hearty and hefty meal could everyone have a sound sleep.

It is a trait of the *Meiteis that they eat till their stomachs are literally full. The Japanese, and for that matter the European's concept of a rich, wholesome and balanced diet loaded with high colorie and protien are of no significance to us. Nothing is more satisfying to a Manipuri than a heavy lunch and dinner of rice, *dal*, fish and *Iromba*. (a traditional Manipuri cuisine of boiled potato and vegetables with lots of chilly and dry fermented fish).

---

* The dominant community living in the valley of Manipur.

3

MITHILA MUSEUM - I

We returned the next day at 2 PM by ship to Mithila Museum which played home to us during our 3 months stay in Japan. Even if we had to leave the place temporarily for a few days for giving performances elsewhere in Japan, we returned invariably to Mithila. I am sure my readers may like to have a little more detailed information about this Museum.

The very name of this Museum must have given an impression to the readers that it is not a Japanese word. Indeed, it is not. It is named after the Indian village Mithila in Bihar's Madhubani district. There are many voluntary organisation in Japan working for the promotion of Indo-Japan friendship and forging a close cultural link between the two countries.

I also met many Japanese who were genuine admirers of Indian Culture. Tokio Hasegawa-san is one of them. He has his own house in the posh locality of Tokyo at the renowned Ginza street. I was given to understand that land in Ginza was the most expensive, not only in Japan, but also in the whole world. About 20 years ago, Hasegawa left his comfortable home in Tokyo and came to Oike village lured by the love of nature. And in the heart of the forests of Oike, he established the Mithila Museum. This is a tiny and sparsely populated village of hardly 15 to 20 families, situated in the Tokamachi city of Niigata Prefecture. This place brought to my mind our hill districts of Ukhrul and Tamenglong. The topographical feature of Japan is very much similar to that of Manipur - full of hills and streams and rivers. The lofty electric transmission towers reminded me of the Loktak Hydro-electric Project and its transmission towers dotted on the hills of Manipur. In terms of weather, climate and topography, there is hardly any difference between Manipur and Japan.

A Madhubani painting by Smt. Godavari Datta

Some years ago Hasegawa-san visited India and familiarised himself with its art, culture and crafts of the land. During the course of his sojourn in India, he came to the Madhubani village of Bihar and was charmed by the Madhubani paintings and its terracota works. He had made up his mind to exhibit the rare folk art forms of India which had never been seen before by the people of Japan. I find in Hasegawa-san something exceptional, even peculiar, from other Japanses people in his personal habits. He is a strict vegetarian. Not only himself, his whole family, including his wife and the two children are pure vegetarians. I did not have any knowledge about his 80 year old mother who was then living with him at the Mithila Museum. There was yet another mild surprise for me. Although he did not even touch either fish or meat, he was fond of alcoholic drinks!

In Japan beer is liberally consumed by both the sexes. I scarcely found anybody drinking plain water during Lunch or Dinner; all of them drink beer in lieu of water.

The distance from Tokamachi to Mithila Museum would approximately be 20 kms. From the broad highway of the city, the road diverted towards the village of Oike.

Both sides of the road leading to the village are covered with thick pine forests - not naturally grown, but painstakingly and meticulously planted by human hands. After a drive of about half an hour, we arrived at our "home" - Mithila Museum. In the vast compound, there are about 5 or 6 houses - all made of wood and C.G.I. shcet-roofed. The twelve male artistes were put up in a Guest House while the seven female artistes stayed in Hasegawa san's two storied building. The Mithila Museum is located just adjacent to his house which had a connecting corridor.

The Museum housed a number of Mithila paintings and terracota idols representing episodes from Hindu mythology. As you are aware, terracotta is made of mud and baked in fire in an oven for the finishing touch. Two terracota artisans, who were brothers (Lala Pandit and his younger brother Jagdish), were already there at the Mithila Museum when we arrived. They would stay at the Museum for another six months, I was told, making terracota figures for the Museum. In the previous year too, they had come to this Museum on assignment from the National Museum, New Delhi and with the active support of Hasegawa-san, it was possible to exhibit a large number and variety of terracotta figures in the Mithila Museum. The two craftsmen-brothers made the clay model of Krishna killing Putana, the demoness of the Hindu mythology. The clay figures were at least $2^1/_2$ feet to 3 feet high. An exceptionally large terracotta was that of a king riding on an elephant. It was almost 4 feet in height. These clay idols are fired in a kiln after they are made and for that purpose a kiln was set up in the Museum complex itself. The firing/baking of the finished clay figures was done by both the Bihari craftsman brothers. After finishing the process, the idols were displayed in the Museum alongside a number of Madhubani paintings.

Groups of visitors, mostly young boys and girls, came to visit the Museum and the crowd was especially large on Sundays & holidays. Each visitor had to buy a ticket costing 500 yen (equivalent to our Rs. 150.00 app).

The Museum Building is located at a higher plain than the surrounding area. Adjacent to that there is a baseball court. Tokio Hasegawa-san is a man of many hues - he is a culture activist as well as a baseball enthusiast. Every evening at about 4 PM, he would come out in the field to play baseball. For practising the game, at least two players are required. While one holds the bat, the other has to bowl the ball towards him. Since Hasegawa-san did not have any extra hand to throw the ball for him, a throwing-machine duplicates the bowler. A brief description of the working of this bowling machine would, perhaps, interest you.

First of all, he puts at least a dozen balls in the machine. An electric motor is then started and the machine automatically throws a ball towards the player at a very high speed. After a few minutes, another ball is thrown out by the machine and the process is continued till the last ball placed inside the machine is exhausted.

The arrival of the artistes from Manipur turned out to be a boon for Hasegawa-san. Whenever we did not have any programme, he started playing with our boys, who bowled for him and retrieved them, too. How pleased he was when our artistes, also acted as amateur bowlers! This was so because in his village there was no surplus man who could afford to lose valuable time bowling for him. In Japan human labour is a scarce commodity; since everyone is busy with their own jobs, Hence, as far as practicable, machines substitute human labour.

At a little lower plain from the baseball ground of Mithila Museum, there exists a low-lying area filled with water. Being a hilly area, there is no dearth of water here. A number of lotus plants were grown, painstakingly and lovingly planted by a Japanese couple. Every morning they came there in their car and tended the lotus plants. With a small manually operated machine they cut the grass and weeds in and around the lotus pond. At the end of their job, they used to pluck three or four full-bloom lotus flowers and returned home in their car. This routine was repeated every morning. I became curious and one fine morning I ventured to ask the couple why they were planting the lotus and tending them so lovingly and painstakingly. "It is for offering to the God", he replied very casually. Everywhere in Japan there are numerous shrines for the followers of Shintoism and Budhist Temples. Like the Meiteis, the Japanese follow a synthesis of two religions viz, the old Shinto religion and Budhism.

In a low-lying part of Mithila Museum there is a big pond. Lots of fishes are reared in this pond - not for human consumption but for the purpose of fishing

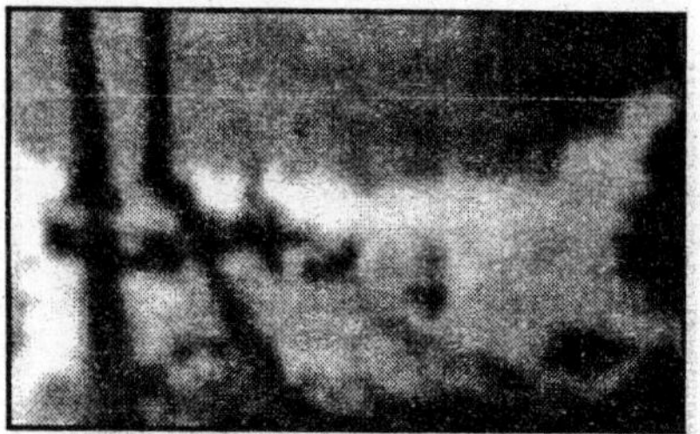

Mithila Museum, Fishing

(mainly angling). The water level of this fishing pond remains full throughout the year; it never dries up. The secret is that it is being fed by perennial streams flowing down from the surrounding hills. In order to maintain the water level of this pond, excess water is drained out on the other side of the pond. There are plenty of fishes in this pond. But they are all unfit for human consumption as they are full of bones. Each and every person who comes to this pond for fishing has to buy a ticket. People started trickling by early morning equipped with their fishing gears and Lunch Boxes. Throughout the day they sit by the bank of the pond hooked to their fishing rods and lines just for the fun of fishing. As soon as a fish is hooked, it is gently lifted and unhooked. Thereafter the fish is again released into the pond.

Their pleasure is only in the excitement and the thrill of catching the fish, not in eating it. In fact, the fishes in the Oike Pond, I am told, are a special species meant only for angling. Adjacent to the pond there stands a Shinto shrine. It is a shrine only in name, but something akin to our clubs. People sit inside it, smoke cigarettes, drink beer and indulge in merry making and loud laughter which again was another culture shock for us. Our concept of a shrine is a sacred place where smoking and drinking is a taboo. We never utilise our shrines in the manner in which the Japanese were doing. May be, this is the culture of the land. In Manipur nowadays, decline and degeneration of cultural traditions have seeped into our society. People smoke Bidis and tobacco during the singing of Harinam Kirtan (devotional song) in religious functions like the Shraddha Ceremony. It reminded me of a sarcastic remark from a friend from New Delhi that Manipur is the only strange place in India where Harinam (literally the name of the lord) and the smoke of Bidi are simultaneously enjoyed by the devotees!

Adjacent to this fishing pond, Hasegawa-san has another Guest House. This is reserved for V.I.P. visitors to Mithila Museum (when Ratan Thiyam visited Mithila, he was lodged in this V.I.P. Guest House). Mithila Museum is situated at a place surrounded by hills. Hilly area no doubt, but it possesses all the modern amenities like electricity, telephone, running water etc. which tends to mitigate the feeling of living in a remote hilly area. Due to heavy snowfall during winter the roads of this place are impassable. When we were in Japan, snowfall had not yet started. However, I could gather an idea of how severe, the snowfall there must be. Almost all the cars in Japan are fitted with Four-wheel Drive. I am told that without 4WD, no car can ply on these roads laden with thick snow. The very sight of 4WD cars in Japan itself was a surprise to us because in Manipur (and for that matter in India) we only have the 4WD fitted in Jeeps and that too for use exclusively in steep hilly roads.

For the first few days of our arrival at the Mithila Museum, we were totally disoriented due to the difference in food and drinks and weather. The Japanese rice contains a very high percentage of starch. The rice after it is cooked, is so soft and white and oily, that it would easily defeat any type of rice found in Manipur. To the Japanese, their rice do not create any problem since they take very little quantity at their Dinner / Lunch. But we, the Manipuris are a strange race and the little lump of rice which the Japanese usually take, would not do for us. And when the oily, starchy and very tasty Japanese rice filled our stomachs to the brim, you can well imagine its effect. Many fell sick with upset stomach. The Japanese rice simply did not agree with the Manipuri stomach. I requested Hasegawa-san to do something about it. Thereafter, he supplied us with Thailand rice which was quite upto our liking and taste. It contains less starch and tastes somewhat like our super-fine or Parmal.

Since the entire family of Hasegawa-san is vegetarian, we were not allowed to cook either fish or meat while staying at Mithila. But eggs? Yes, to eggs he had no objections. Prolonged and continuous eating of vegetarian foods since our arrival in Mithila had adversely affected the health and morale of our boys and girls. Manipuris are traditionally fish eaters. The effect was more pronounced in respect of our Pung-cholom and Thang-Ta artistes. Their performances involve strenuous movements and heavy physical exercises. Needless to say, they require high calory protein diet and when they were treated only with rice and *dal* (*lentil*) after a hectic performance, you can well imagine the consequences. The artistes started grumbling. I made it known to Hasegawa-san about this development in our camp. At first he was not very accommodative, but after some discussion he relented. We arrived at a solution that there was to be no fish or meat while staying at Mithila, but outside it there would be no restrictions. Cooking of all our meals was done by our girls by turn since that suited our tastes best. As for the provisions like rice, *dal*, vegetables, fish, meat, eggs etc. Hasegawa-san provided them in plenty.

4

SENDAI CITY

On July 26, we left Mithila for Sendai City - a drive of nearly 8 hours. Our air-conditioned micro-bus sped at the steady speed of 120 kms. per hour. We passed on the wide road running through the hills - hills in front, hills to the left and right. The roads run through tunnels; they never take round the hills; the vehicles drive straight into the bosom of the hill through innumerable wide tunnels, some of them as long as 10 kms. Those tunnels have two lanes to facilitate two vehicles to pass each other. The interior of the tunnels are not at all dark. It is fitted with powerful electric lights and exhaust fans for sucking out the foul air inside. There was no human being in sight inside these tunnels. Robots holding flags in their hands give traffic signals to the passing vehicles. Countless

Traffic Police Robot

number of vehicles ply on the Highways but none of them use horns; none emit exhaust fumes. Among the trucks plying on the wide metalled roads of the Highways of Japan, there are some realy huge ones having as many as 14 wheels and as big as our Mamang Sangoi (outhouse of the Manipuris). To us whose concept of a big truck is limited to a TATA *Baro-Sou-Dus* (Tata, 1210) and ASHOK LEYLAND, the very sight of these huge Japanese trucks itself was a thrill and evoked a mild surprise.

Although not a single soul was in sight throughout the long drive on the highways, telephone boxes were installed by the side of the roads at intervals of 10 to 15 Kms to facilitate informing the police in case of road accidents on these long unmanned stretches.

Starting from 7 AM we stopped at a Parking Area at 1 PM for our Lunch. There are such parking lots for every 40-50 Kms on the Highways which provide all the comforts and facilities to the travellers such as telephone, eating,-houses and restaurants, gift shops, bathroom, toilets etc. Because of these excellent facilities available on the Highways, driving 1000 Kms a day in Japan does not cause any hardship and inconvenience to the travellers. All the artistes were supplied with one packet each for lunch. The packet contained noodles and we were initiated to the habit of using chopsticks, because in Japan nobody eats with their fingers, as we do in India. Everybody eats with the help of chopsticks. Initially we found it quite difficult to use the chopsticks but gradually we were adapted to the habit. The parking lots were provided with automatic water-taps; you simply push your hand just below the tap and water flows out of the tap automatically. The moment you remove your hands from the taps, the water stops running instantly and automatically.

On arrival at Sendai, we were lodged at the Youth Cultural Centre. We had to give our performance there for two consecutive nights. We found the arrangements at Sendai a little inconvenient in that the venue of our performance was at one place while arrangement for our food was made at another. For our stay during the night, we had to drive for one hour to a Forest Lodge in the hills. In Japan accommodation is very expensive; in fact, everything is expensive in Japan. That Forest Lodge where we stayed belongs to a friend of Hasegawa-san. When this was explained to us, we gladly co-operated with him. We were motivated solely by the thought that, inspite of these minor irritants, we strongly felt that we had a mission to expose the culture of Manipur to Japan.

Along with our performance, Hasegawa-san mounted an exhibition of the terracottas of the Mithila Museum. At night, cultural items of Manipur were performed while in the day time, the terracottas were exhibited. On the day of our performance, during the sword-fight by our Thang-Ta artistes, one sword broke and the piece went flying. Mercifully, no one in the audience was hurt by the flying deadly missile. At the end of our Naga-Dance, one of our female artistes collapsed on the stage due to over-exhaustion. But on the whole, the performance was a success. Each and everybody who came to witness our show had to buy a ticket costing between 2500 to 3000 Yen (equivalent of about Rs. 1000/-). The stage where we performed was quite big and spacious. The powerful lighting system of the hall contributed not the least, to the quality of our performance.

After the performance, we left for the Forest Lodge for staying the night. It took nearly one hour by Bus. On reaching there, our girls started cooking dinner. A very memorable incident of that night's dinner was the surprise menu of 'Laphu-Iromba' (a Manipuri delicacy prepared with banana-stem, potato, red chilly and dry fermented fish). How come 'Laphu-Iromba' in Japan? Wherefrom was the 'Laphu' (banana-stem) procured in Sendai city was a mystery to me. Later, it was learnt that the 'Laphu' of the Welcome Gate of the morning function was retrieved by one of our resourceful boys when the function ended. But where on earth could they procure 'Utonga' (fermented fish which is the most essential ingredient for preparing 'Iromba') ? That too was smuggled into Japan by our enterprising and far sighted boys !

I cannot help but admire the ingenuity and courage of our boys who could manage to pass the 'Utonga' through the tight Security of Japanese Airport where strict vigilance was maintained to check the entry of such prohibited food items into Japan. That was the first night when we partook 'Iromba' in Japan and it seemed to taste heavenly! May be, we were deprived of this unique Manipuri delicacy with a top-dressing of " TONINGKOK " (a highly aromatic evergreen herb, usually taken with 'Laphu Iromba' in Manipur). Though plenty of Toningkok are found growing wild in Japan, the Japanese do not eat it. They probably are not aware that this green leafy vegetable is eatable.

5

YOKOHAMA - 1

The next day on July 30, we left Sendai City for Yokohama at 9 in the morning. After a drive of about 8 hours, passing through uninhabited stretches of land and countless number of tunnels, we arrived at our destination at 5 in the evening. Arrangements for our stay was made at a Children's K.G. school. The next day the Manipur Cultural Troupe was Scheduled to perform in the "Yokohama Summer Night Fest '95".

As is my daily habit of going for a morning stroll, I went to a near-by Park, accompanied by two young Pung-Cholom artistes. This Park, originally developed as a Horse Race Track sometime in 1930, is now converted into a Public Park. Just opposite to it, there exists a vast complex of the U.S. Navy Fleet. This was quite a surprise for me to find a foreign military base in Japan. I fail to understand how they reconciled the sovereignty of their nation with a foreign base. In India, we do not entertain foreign Military Base on our soil.

Yokohama, U.S. Navy Fleet

There were lots of Americans there. I met two U.S. Marines and had a chat with them. It was a pleasant and memorable meeting, not for anything special, but for the simple pleasure of meeting an English speaking person. In Japan very few people speak English. We speak to them, of course in English, which they don't understand. They, too, speak to us, obviously in Japanese language, which unfortunately we don't understand. Language is a great problem for a visitor in Japan; it is hard to communicate with them. That was why an English speaking interpreter was invariably present wherever we travelled in Japan.

In the same Park where I went for the morning stroll and jogging, there were a number of Japanese morning walkers. One thing that struck me was that none of them were young. Almost all of them were old men and women between the age of 70-80 years. The old women wear long pants and hats on their heads, and took to brisk walking. The Japanese are very health conscious and took good care to remain free from any ailments.

Most of these early morning birds bring their own pet dogs. Keeping pets is also a national trait of the Japanese people. There is a large population of old and aged persons in Japan. Wherever we went, we came across persons, majority of whom were old and aged. Most of the people who came to witness our performances were also of this advanced age group. Such a sight of the old and aged persons made me ruminate. The high rate of life expectancy in Japan has the potential of causing social problems in the long run. It will be a nightmarish experience: if people simply refuse to die, what would be the shape of that society? I simply fail to imagine. According to a latest report, there are more than 7000 Centenarians in Japan and the number of such people are on the rise every year. Among the hundred year-olds, 80% of them are women. Japan has the record of highest life expectancy rate in the world. In this country, the aged and the old people were accorded with very respectable status. Every year September 15, is being observed in Japan as 'Aged-Day'. The average longevity of a male in Japan is 77 years while that of a female is 83 years.

I was curious to know the reasons behind this. It can be attributed, not to one but to a varied reasons:

(i) It can safely be stated that Japan is almost a pollution free country. Breathing the clean air free from dust and poisonous fumes is the most significant factor. As had already been mentioned, no vehicles emitting exhaust fumes ply in the Japanese streets.

(ii) Secondly, active and excellent health care measures are taken up by the government to maintain the health of the citizens. When a person reaches the age of 80, it is the responsibility of the state to cover all risks by providing Health Care Insurance. This ensures free treatment in Hospitals including medicine when the old person falls ill. Expenses for medicine (upto 30% of the cost) are re-imbursed by the state. This itself is one of the major contributing factor for the health and long life of the Senior Citizens of Japan. Japan is a very affluent country; it has a rich economy and food and drink are bountiful. The very fact that they eat healthy and nourishing food is another factor.

Yokohama is a Port City. The venue where we would be giving our performance was a vast expanse by the sea-beach. An open-air stage was constructed against the backdrop of the ocean. A memorable incident happened on that day when we were giving our performances. The Pena (indigenous stringed musical instrument of the Meiteis) which we brought from Imphal had been left behind at Calcutta Airport inadvertently. Our request to trace it and despatch it to Japan had not materialised. I was at a loss. Wherefrom could I obtain a new Pena in Japan? But performing 'Lai Haraoba' dance without a Pena is unthinkable. Yet we were bound to perform, since it had already been scheduled and was now impossible to postpone. In such a situation, the only alternative was to make a look-alike of the instrument at Yokohama. That very night we improvised and prepared a dummy Pena with whatever materials were available With this dummy Pena, we began our performance of Lai Ikouba and Lai Haraoba. The Maibis chanted oracles before the deity; the Amaiba invoking theYakairaol to the accompaniment of Pena. Our 'Pena' looked exactly like the real one but was incapable of producing any note. Instead of horse-hair, we set only a thin copper wire. The performance ended successfully and we congratulated each other on the success of the 'Pena'. We secretly enjoyed the fact that we had taken the Japanese audience for a ride. But our joy was short-lived. Soon streams of Japanese from among the audience came to confront us back-stage. "How is it that your indigenous musical instrument does not produce any sound?". When I narrated the saga of our lost Pena at Calcutta Airport and how we had to prepare a dummy overnight, they left laughing graciously.

That night another memorable incident happened at the sea-beach. We were desperately searching for a few flowers for our customary ritual of stage-puja before each performance. There were lots of flower-beds and plenty of flowers blooming nearby. But we had been dissuaded from plucking even one flower.

We were told that in Japan nobody plucks flowers like that. The organisers were kind enough to buy some flowers and handed them over to us for the Puja. Among the crowd who came to witness our performance there were hardly any Indian present. Most of them constituted of American and European tourists.

6

BATO-CHO

On 1st August at 8-30 AM we left Yokohama and reached the small town called Bato-Cho. This place is situated deep inside the hills. The name Bato-Cho comes from the Chinese word 'Ba' which means horse and 'To' means head- Bato-Cho, Horse-head. This town is named after a Buddhist icon with the head of a horse called Bato-Kannon. In Japan, there is hardly any difference between a small town and a big city in terms of infrastructure and modern day facilities. In this small town situated in the interior hill area, there is nothing which is lacking such as good road, regular supply of electricity and water, Telephone, Fax and STD/ ISTD facilities etc. Thus, we did not have the slightest feeling or handicap of staying in a remote place far away from the city. Both sides of the roads of Bato are covered with green forests of dense Pine trees, meticulously planted in rows by human hands. Driving through the green paddy fields,

Bato-Kannon

we could see from our bus vegetable gardens with varieties of fresh vegetables like egg-fruit, onion (Japanese onions are as big as a tennis ball), cabbages, soya-bean, etc. On the slopes of the hills are grown American corn (maize). The main vocation of the inhabitants of this place is agriculture.

When our party arrived there, a large number of women were waiting for us. They were keen to learn the Indian style of cooking. They requested our girls to join them in cooking dinner for that night so that they

Bato - Town

could learn something from them. Immediately I went to the shopping centre of Bato and purchased fish and other provisions required for the dinner. Our girls together with the Japanese women entered the kitchen and they started cooking. Many of the Japanese women held note books in their hands and asked questions to our girls and the points were jotted down. They showed much interest and keenness to learn the intricacies of Indian cooking. Our girls did not cook anything very special except fish, *Dal*, *Subzi* and *Iromba*, yet everything, seemed quite a new experience to the Japanese women. Our style of cooking with lots of spices and slightly hot with chillies was to them an entirely exotic fare and they ate it with real relish. It was quite a fun watching the 33 Japanese women with tears in their eyes (due to the hot chilly) joining us in the dinner. Some of them even attempted to copy our style of eating with our fingers. They, too, started eating like us but soon changed to their own way of eating with chopsticks. They liked the dinner and appreciated it, not because it was very tasty, but simply because it was totally a new fare. The different food items served in a Japanese Lunch or Dinner are not suited to the Indian habit of eating rice and curry. For the cooking medium, they never use mustard oil which we prefer. Cooking meat or fish in Soya oil is never the same with our cooking in mustard oil. Mention may be made here that almost all the varieties of Indian spices are also available in Japan - *Hing* (asafoetida), *Methi* (fenugreek), *Long* (clove), *Jeera* (cumin seeds), *Elaichi* (cardamom), *Pappad*, dry-fish, *Hawai-char* (fermented soya-bean), Bamboo-shoot, etc. They too, eat these things; the only difference is in the style and method of cooking.

There was a social gathering after the dinner. Two Japanese youths came forward and displayed their Sumo Wrestling. We reciprocated the gesture by displaying Manipuri Mukna (traditional wrestling of Manipur) by two of our artistes. Our Japanese hosts were highly pleased to watch our Mukna demonstration which has close affinity with the Japanese tradition and culture. Next, the Manipuri girls presented a Japanese song (they learnt it within a few days of our arrival in Japan) - SIYA WASE... NANA TERA... WATA KO... This unexpected Japanese song by the Manipuri girls was a pleasant surprise to our hosts; they smiled and applauded our artistes. The social gathering continued till late in the night and cemented our friendship and affection with the Japanese.

The next day we presented our cultural programme. The auditorium was not very big; it could accomode about 300 people. Yet it can safely be concluded that almost the entire population of the small town had turned-up to witness our performance. Among those people who came to watch our dances, I met one American girl. I was a bit surprised to find a lone American girl in such an isolated hilly place. I was given to understand that she was on an American government scholarship to work in the villages. Another very pleasant surprise for me was a meeting with a Japanese girl named Kumiko Ohno. In fact, she came to meet us when she learnt that a cultural troupe from Manipur were performing at Bato-cho. She told me that her best friend had been married to a Manipuri. We never imagined that a girl from faraway Japan had married a boy from Manipur and settled there. She gave us the Imphal address of her friend and requested us to convey her love and well-wishes to her friend. What a touching gesture! I assured her that I would most certainly hunt for her friend on my return to Imphal. She then wrote down the name and address of her friend in my diary :

Mrs. Yoshika Ruivah
Chan Ngam House,
Fairyland; Stadium Road,
Imphal, Manipur

This girl Yoshika is married to one Tankhul Youth named Ramchan of Ukhrul, now settled at Imphal. While studying Agriculture in Tokyo, Ramchan fell in love with this Japanese girl whose parents gave their consent and blessings to their marriage. After their marriage the girl came to Manipur with her husband. A few days after my arrival from Japan, I went in search of the whereabout of Mrs. Yoshika Ruivah. Whithout much difficulty I could find the couple and met them in their home at Stadium Road, Imphal. How pleased they were and what a fulfilling experience for me! It seems Japan and Manipur have come much closer, all of a sudden.

7

ONIKOBE

On August 3, we left Bato-cho and arrived at Naruko Town in Miyagi Prefecture. It was about 6 hours drive and we covered around 500 kms. The topography and landscape in Japan is almost the same everywhere - high hills, thick jungle, dense pine forests etc. We drove on the wide metalled road in the hills and reached Naruko Town. It is also a town situated deep in the hilly area.

We gave our performance at a place called ONIKOBE in Naruko Town. This is a small village having only about 500 families with a total total population not exceeding 2000. It is located at the foot of a high mountain. There was heavy rain on the day of our arrival and the temperature suddenly dropped down. I needed wollen socks and went to buy a pair from a nearby Departmental Store; it cost me 350 yen (our equivalent of Rs. 100/-). Things are very expensive in Japan - poor Indians like us could hardly afford it. I would have more occasions in subsequnt pages to show how expensive a place Japan is. Although Onikobe is a sparsely populated village, it is one of the most popular Ski Resorts of Japan. During winter, the surrounding high mountains are covered with snow and ice and thousands of sportsmen converge in this small village for skiing. Naruko Town is also famous for its hot springs which are very popular with the tourists.

For a single performance of our Dance at such a nondescript and isolated place, it was simply baffling to us as to how much expenditure they had incurred for it. One obvious reason may be because the Japanese are affluent people and they have lots of money to spend. But there is yet a deeper and a more significant reason. I am grateful to Hasegawa-san for enlightening it to me. When I asked him the reason why had he brought us to that small and insignificant village, his reply was both revealing and inspiring. It explains the whole concept and perception of culture in Japan.

According to him, culture is not the monopoly of the towns people only. People living in the remote villages, have an equal right to have their due share. Those who live in big cities have ample opportunities for exposure to different art forms from different countries whilst people living in such remote villages have been deprived of the same. They too, deserve to be treated equally.

Hot springs are abundant in Japan since the country has many volcanoes. We had our first experience of taking bath in a hot-water spring during our stay at Bato. The bath house was well constructed. Inside, there exists a shallow cemented Pond into which hot-water brought from the hills by pipe is released. The Pond is full to the brim with the hot water from the spring, containing minerals and sulphur, beleived to possess healing properties for arthritis etc. Hot-water baths are very popular with the Japanese; and they indulge for hours immersed in the hot-water Pond.

These baths are divided into two seperate portions - one for men and other exclusively for women. Every body had to buy a ticket for using these bath. On the first day when we went for the bath, we had the shock of our life! No one could enter these hot-water ponds with their clothes on. Prior to entering the pond, one has to strip-down to one's skin and then enter the bath stark naked. Not even a scanty piece of cloth in the form of a panty or an underwear is pemissible. How can we behave in this childish manner at this ripe old age?! I was caught in the Hamletian dilemma "To strip or not to strip". While we were hesitating like this, a group of Japanese entered the cloak-room, removed their garments without any inhibitions and proceeded stark-naked straight into the hot-water pond. In this place there is no "Q" system - everyone enters the pond without waiting for others to come out. This emboldened us and like a child with faltering steps, we joined the Japanese bathers. Although a little uneasy and hesitant initially, it was not long before we all had completely been adapted to this quaint Japanese way of life.

Bato, Hot-water Pond

Onikobe, being a small village with no proper auditorium, we gave our performance in an open lawn. It is customary in Japan that everyone lend a helping hand in organising a public function - before its beginning and after the end. Everyone helped to clean the venue of the performance by removing the chairs and dismantling the lighting equipments etc. When our performance was over, no one left the place without helping the organisers; some folded the chairs, some people dismantled the P.A. system and the lights and loaded them in a small pick-up van and drove away instantly. After that, every one joined to remove the litter on the ground and swept the area clean. In this way, there was no tale-tell sign left that only a few minutes ago, the place was a performing area. This lovely habit of community service of the Japanese deeply impressed and moved us. There is no coolie system in Japan, as distinct from India; everyone has to do it by themself. This type of manual labour or self help is much valued and appreciated in Japan. This had brought a contrasting image in my mind when I organised such functions in Manipur. You can't dream of such help and co-operations from the spectators. Everybody would leave the area littered with pieces of paper, polythene bags and eatables, etc., quite unconcerned. It is always the responsibility of the poor organiser, who have no other alternative but to hire a few hands from "Shamu Makhong" (an euphemism for labour Bazar).

This commendable system of the Japanese impressed and inspired us so much so that we also joined them in cleaning the area and loading the chairs in the van. And how grateful were our Japanese hosts who thanked us profusely. Onikobe is renowned for its supply of good timber–most of their houses are built of wood. Its wooden constructions

Wooden House at Onikobe

and structures have earned a name in Japan. The huge Town-Hall of Onikobe, constructed entirely of wood without using any iron-girders is a testimony of their skill in wood-works.

8

IWAIJUMI

On August 4, we left Onikobe at 10 a.m. and reached Iwaijumi at 3 p.m. It is also a small town in an interior hill area having a population of about 4500 only. For miles, the area is uninhabited and the vast stretch of road is so wide and excellent that we found hardly any difference between a city and rural area. Mrs Nabuko Sasaki had been waiting to receive our group. We had already been introduced to her and known each other at Sendai. She had come all the way from Iwaijumi to Sendai just to have a pre-view of our programmes there. She is an amazing lady, with an unlimited energy and charm. An activist in the field of arts and culture, she was the Chief Organiser of our performances in Iwaijumi.She was all smiles when she saw us and made arrangements to make our stay for the night comfortable.

Before the start of our performance, the children of Iwaijumi presented a 30-minute cultural programme. They started the programme by singing the National Anthem of India. This was not only a surprise, but also quite a thrill when unexpectedly and unannounced, the children of Iwaijumi sang in chorus "Jana Gana Mana

Adhinayaka Jayahe Bharata Bhagya Vidhata". All the Manipuris stood up as a mark of respect to our National Anthem. Strictly speaking, as per protocol, it was not correct to sing the National Anthem for a small Indian group such as ours. But the spontaneous outpouring of our National Anthem from the Japanese children in this remote town was considered a tribute, not to the Manipuri Troupe, but to India. I take it as a gesture of mutual respect and goodwill between these two countries.

Our performing space at Iwaijumi was neither an auditorium nor a Theatre Hall; it was a large Gymnasium (wherever we visited in Japan, Gymnasiums abound). It had a wooden floor. All the spectators removed their footwear before entering the Gym. Girl volunteers at the entrance provide one cellophane bag to each of the spectators who put their footwear into it. The disciplined spectators entered the Gym and seated themselves in neat rows on the Tatami spread on the plank floor; there were no chairs. In a town of about 4500 people, a thousand people came to witness our performance and jam-packed the entire Gym. Each and every item of our Manipuri Dance were lustily applauded and cheered by the enthusiastic audience. There is much similarity between the culture of Manipur and Japan, about which I shall have more opportunity to narrate in the subsequent chapters.

After the performance, we gave a curtain call to the appreciative audience and offered our customary salute to the audience by bowing our heads in unison. The Japanese were highly impressed. They presented us with bouquets and small gifts on the stage. We reciprocated the gesture from our side by presenting a "Khamba-

Thoibi" doll to the Chief Organiser Mrs. Nabuko Sasaki. How pleased and grateful they were to receive the present, a small token of love from the people of Manipur. They thanked us profusely.

One special quality of Mrs. Nabuko Sasaki is that she speaks English well. This made us quite comfortable while communicating with her. An unforgettable incident at Iwaijumi was that four Japanese girls befriended our girls. They were not separated even for a moment from our girls. I do not know what unseen force had attracted them to each other. Their love was infectious. I wonder, as to what had prompted so much love and camaraderie between these two nationalities. It shall forever be etched in my mind the touching spectacle when the Japanese girls sobbed at the moment of our departure from Iwaijumi. Our girls also cried and waved their hands from inside the bus. The Japanese have close resemblance with the Manipuris in their facial features and general appearances. May be because of this trait, they showed unusual friendship and affection for us and made us feel like their own kind. We, too, never for a moment felt that we were in a different country amidst alien people. The love and affection they bestowed on us, their warmth of feeling towards the Meiteis did not bear the slightest tinge of pretension. Their smiles were not the cosmetic smile as an act of formality and normal courtesy. Their hospitality exude an invisible message of genuine love and friendship which we felt intuitively.

Iwaijumi

That night we were lodged in a big air-conditioned Hall. Next day, we would be shifting to another accomodation. In Japan accomodation is always a problem and I have personal knowledge of the organiser's difficulties in accomodating such a large group of dancers from Manipur in so small a town. They lodged us in whatever places were convenient to them and we never raised any objection to them for minor irritants. Precisely because of this mutual understanding, we were able to stay for such a long period in Japan giving more than 40 programmes in different locations. It will not be an exaggeration to say that no other cultural organisation can independently of their own venture into such a long trip to Japan without the support of local voluntary organisations. Imagine the expenses of 19 artistes for three months in Japan - their food and lodging, travel expenses and the to and fro Air-fare between India and Japan, etc., all these put together would amount to a few crores of Rupees.

Riusendo Cavern

One morning we were taken out for sight-seeing to the Ruisendo-Cave. It is a magnificent cave inside the hills. Once we entered the cave, we could see the clear blue underground streams and vast water bodies. The interior of the cave was quite cold (22 degree celsius) compared to the outside temperature. On the day of our visit to the cave, there was a heavy downpour that worsened the weather and it became much more colder than usual. Since we had not taken any warm clothings, we felt much discomfort inside the cave. There were iron railings all along the passage inside to protect the visiting tourists from falling down into the abyss of the cave. It was indeed a wondrous sight which we had never seen before. As already mentioned, that night arrangements for our stay was made at a different place. It was a Farm House located at the foot of a hill, about half an hour's drive from our place. It appeared that the said Farm House had been left unoccupied for a long time. The rooms were full of cobwebs when we entered the house. But the most surprising fact is that the Farm House was newly constructed and abandoned there for any eventual use. Only occasionally it was occupied by visitors and tourists. Although left unoccupied and without even a caretaker to look after it, all the fittings and furnishings inside the Farm House remained intact and nobody stole anything. As we entered the rooms, we found that the Farm House was fully equipped with all facilities such as Electricity, Running water, Telephone, Television, Refrigerator, Cooking gas, Utensils and Crockeries, etc., everything that a visitor or tourist would need for his stay there. There were a number of such Farm Houses constructed in the foot hills where anybody can stay on payment. The occupant had to pay room-rent and other charges before leaving the house. There is no one to collect the rent - you have to deposit the same on your own in a cash collection Box placed there. When we opened the refrigerator, it was filled with liquor, Beer and Cold-drinks. You can take anything you like but not free of cost. You have to pay for it by depositing the cost of the drink in the Collection Box. What a lofty ideal of public morality and civic sense! Can you dream of such a Farm House in Manipur? It won't work here; everything inside would be stolen.

Being surrounded by high hills on all sides, the Forest Lodge was quite cold. Not very far from it, there is a Dairy Farm called Nakahora Dairy Farm. Having nothing to do, we went for a visit to the nearby Dairy to while away our time. I had a chat with the owner of the Farm. He informed us that all his cows were let loose in the hills for grazing the natural fodder. Milk from such cows are known as "Environmental milk" and there is great demand in Japan for such type of milk. This is supposed to be special kind of milk from cows reared, withhout any artificial feeds. The owner presented us with a good quantity of this milk. When we boiled and tasted it, we found it almost the same with the normal milk. Perhaps, it was only a psychological satisfaction that the milk is from cows eating natural fodder.

On return to our Forest Lodge we found a large quantity of fish sent by Mrs. Nabuko Sasaki who knew the Manipuris' weakness for fish. But those fishes lovingly sent by her were all sea fishes having pointed heads resembling snakes. The very sight of the fishes was so repulsive that everyone refused to eat it and they had to be discarded. The next day early in the morning Mrs. Sasaki came with her daughter driving a car herself and bade us farewell. The four Japanese girls (about whom I had mentioned earlier) - Sachiko, Noriko, Fouko and Miyuki also turned up just at the moment. They and our girls fell on each other's arms and gave a tearful farewell. It was indeed, a very touching scene.

It was 9 a.m. All of us boarded our air-conditioned Toyota micro bus. Farewell Iwaijumi; Sayonara Nabuko-san and all the lovely Japanese girls of Jwaijumi. We headed for our 'Home'–the Mithila Museum.

Girls of Iwaijumi in Manipuri attire, (back row) Sachiko, Nariko, Fouko and Miyuki.

9

MITHILA - II

Iwaijumi to Mithila Museum is a 12 hours drive. Starting at 9 a.m., We reached our destination at 9 p.m. As advance information of our arrival was sent, a hot dinner awaited us. Mrs. Hasunama, an official working in the Mithila Museum, is well acquainted with Indian Culture and Cuisines, having stayed a long time in Madhubani village of Bihar. She speaks Hindi well and knows Indian style of cooking. On that night she cooked *Arhar Dal* and vegetables *subzi*. The two terracotta artiste brothers, Jagdish and Lala Pandit, from Bihar also lent a helping hand in the cooking.

Our next performance in Japan was scheduled only on 13 August and we had to stay at Mithila for four days. Since we had nothing to do during those days, it provided me an opportunity to study their life and social customs. I considered the opportunity heaven–sent since I hardly got any time to do so amidst the hectic cultural programmes.

The Japanese are well-to-do people and possess high spending power. Wherever I visited, I hardly came across any poor man. The policy of the government ensures to maintain an equilibrium between the rich and the poor. In terms of financial security and material possessions, there is hardly any difference between a highly paid Government servant or a Corporate Executive and that of a simple farmer in a far off village. A mere cultivator also possesses the same amenities of modern-living like a car, Television, Telephone, Refrigerator, etc., just like any other highly paid government employee. It will be pertinent to mention here as to how can they maintain such a happy balance between different strata of society.

In Japan things are very expensive in general. The price of rice is quite high. The rice produced by the farmer has to be purchased at a very high rate by the salaried people. One piece of maize (American corn) costs 100 yen; a pack of 3 brinjals costs 100 yen; a pack of only 4 tomatoes costs 100 yen (equivalent of our Rs. 33/-), etc. In Japan one need not necessarily be in Government Service and earn a monthly salary; it is enough for a person if he plants some cabbages, tomatoes, potatoes, egg-fruit, etc. The sale proceeds from his agricultural products more than meet the expenses to buy all the material comforts which a salaried person can afford.

Massive office buildings, factories and industrial establishments have no chowkidars (night watchman). At the stroke of five in the evening, all the employees of the office, close all the doors and windows, switch off the lights and fans of their offices before leaving for their homes. The next morning, the same routine is repeated in reverse. At exactly 10'O clock (the Japanese are renowned for their punctuality), they would come to office, open the doors and windows and occupy their seats and start work. In Japan 10'O clock means 10'O clock; even a minute's delay is not tolerable. I wonder whether it may be possible in my place to leave a huge building full of valuable and costly items unguarded at night. Being a bit baffled, I asked my Japanese friends the reason, and their reply was "there is no small thief in Japan".

Wherever we travelled in Japan, there were many huge junk-yards full of cars, TV sets, Tape recorder, small earth-tillers and tractors, etc. Having nothing to do, our boys had been fiddling with those cast away items in the junk. To us, they were as good as the new ones. But with the slightest defect in the machines, they discard and dump them in the Junk-yard. This is because they cannot afford to waste their valuable time in mending machines. Sometimes repairing a machine, I was told, is more expensive; it is more economical to buy a new car rather than spending money on the repair of an old one.

They are affluent people who have lots of wealth but I simply fail to comprehend it. My car is not less than 30 years old and I am still using it. Those cars discarded as a Junk in Japan are, I must admit, ten times in better condition and shape then that of mine. Wealthy persons have constructed country houses, far from the madding crowd. They occasionally used them for staying on week-ends and leave them unguarded for the rest of the week. Those country houses are well equipped with many valuable items like TV sets, telephone, refrigerators, gas stove, cooking utensils, bed rolls and other clothing. Nobody ever touched them, not to speak of stealing them. I was amazed and asked Hasegawa-san whether there was no thief around. He was quite amused at my question. He told me, "Everyone has got everything; so where is the need for stealing?" Indeed, it is a Paradise on Earth.

In Japan dwelling houses are by and large, small in sizes. One of the most important reason is that the price of land is quite high. Since the supply of wood is abundant, most houses are constructed of wood; The floors are made of planks and roofs are of corrugated Iron sheets. Invariably, in front of every house, a small garden is laid. Even if there is practically no space for a garden, a few flower pots would surely decorate their court yards. A single pine tree - its branches meticulously manicured, would often be planted in front of their houses. The Japanese possess a high sense of aesthetics. This itself is contrary to our pre-conceived notion of the people of Japan. Not only do they plant flowers in their own homes, they do so even on both sides of their Express-ways for hundreds of miles. And mind you, they are planted not by the government, but on the voluntary initiative of its citizens. This itself was quite a big pleasant surprise for us.

Writing of flowers, I am reminded of another fact. In our country, lotus plants grow wild in big lakes and marshy wetlands. Nobody takes the trouble of cultivating lotus plants. But in Japan there are large patches of land under lotus cultivation with hundreds and thousands of the flower blooming. They were meant for selling in the market. Like the Manipuris, the Japanese people, too, eat lotus-stem (thambow, in Manipur language). I have seen it being sold in many departmental stores and shopping complexes.

10

MATSUYAMA

On August 13, we left Mithila for Matsuyama which is a small town in Miyagi Prefecture. It took eight hours for our bus to reach there. Driving on the Japanese Express ways was a new experience for us. All the cars plying on the street were broadly of two types - one is small like our Maruti cars. The other one is a big family car as big as our Mini-bus. Most of the cars are being fitted with four wheel drive. Whole families move in it for sight seeing or spending week ends. All the vehicles ply in neat orders; none attempted to overtake the vehicle in front; none used horn and the vehicles emitted no exhaust fumes. Although the average speed of the cars on the road are between 120-150 kms per hour, nobody drove rashly. Such huge machines are rare in India because we simply don't have suitable roads to accommodate such big trucks.

On the evening of our arrival at Matsuyama, the townspeople organised a reception for us. The house earmarked for our stay in this town was found to be too small to accomodate our group and caused some inconveniences. But we had no complaint and took it sportingly. The next day we were shifted to a better accomodation. The day of our performance at Matsuyama coincided with a local festival there. It was arranged like a big fair. There were a number of stalls and eating junkets. Hasegawasan too, opened his own stall in which his young son Munna (7 years old) and his daughter Raina (14 years) lent their help. His stall specialised in the sale of Indian goods like Salwar-Kamej Suits, plastic bangles, brass icons of Hindu gods, hand bags, Bindi, Henna etc. Japanese women and girls crowded his stall and purchased the novelty items. The sight was very amusing to us. It was a great fun to watch the Japanese girls trying the cheap bangles on their hands alternately removing it, whispering something among themselves and giggling. A pair of bangles made of ordinary lac, which often we find in Delhi's Janpath, was sold for about Rs. 100/-. Japanese girls do not have the tradition of wearing bangles as ornamants. However, it seems that the love for beauty and beauty-aids by girls of all nationalities, is a universal trait. The girls purchased the bangles, instantly put it on their wrists and departed from the stall.

Wherever we performed, one such stall selling Indian traditional items was invariably opened close-by the venue. During the 15 minutes interval of our performance, the spectators would throng the stall and there was a brisk sale of those Indian goods. The man who was responsible for opening the stall was one Mamoru Miura-san. He was a Judo Instructor. He had stayed for about 10 years in India as Judo Instructor in the National Institute of Sports (NIS), Patiala. He recalled that while in the NIS, Patiala, he had imparted training to one girl Judoka from Manipur (Miss Sabitri Chanu). Because of his prolonged stay in India, he knew our country well and spoke reasonably good English. He had a number of friends in Delhi and Punjab who regularly despatched these Indian wares to him. He also used to undertake shopping trips to India at least twice a year. The items were not expensive ones - only the glittering cheap stuff which are sold on the footpaths of Delhi. But to the Japanese, they are quite a novelty and he sold them like hot cakes during each of our performances. To us who were aware of the correct price of these materials, it was so amusing to see an ordinary Salwar-Kameej costing around Rs. 100-200/- being sold in Japan for a whopping Rs. 1000/- which were hot favourites for the fashion-conscious Japanese girls. Nobody seemed to bother about the price. If they had a fancy for the piece, they just paid fot it and departed smiling, extremely pleased at the possession of an exotic piece from India! What had delighted me most was the opening of stalls selling Indian goods; it complemented our performances by creating an Indian atmosphere and ambience. For a moment, it made me forget that we were in Japan and fancied that we were actually in an Indian city.

By 5 in the evening, all the streets of Matsuyama were a sea of humanity. A procession led by four motorbike riders was approaching. As requested by the organisers, two of our artistes, one boy and a girl, in their dance costumes had joined the said procession. They were seated in the side cars of two lead-Motorbikes. Little children, apparently belonging to different organisations, carried small palanquins in that procession and wended through the roads of the town until it finally converged in front of the Matsuyama Museum. The dais for performance of our programme was also constructed there.

A very memorable incident was that our friend Mrs. Nobuko Sasaki of Iwaijumi came with her daughter to meet us. She brought eatables as presents for the artistes and told me that she and her daughter had come all the way from Iwaijumi, driving for five hours, just to watch our performance again. We feel honoured and humbled. This indeed, was how the people of Japan had appreciated Manipuri Dance.

After the performance, we reached our place for night-stay, a sports Hostel, at 10 p.m. It was almost midnight when we finished cooking and eating our dinner. In most places, we often cooked our own food. Their food does not include our favourite items of rice and curry. It consists of something of everything such as a piece of fish, some soyabean preparations, some green salad and sea-weeds in small quantities. Those items, although rich in vitamins and calorie contents, protein and minerals, never gave satisfaction to the Manipuri palate. Only when we cook ourselves our favourite dishes of rice, fish, dal (lentils) and vegetables subzi, do we really enjoy a meal. Anything other than these meant total disaster during our performance!

I vividly remember the next day; it was the 15th of August - our Independance Day.

As a habit, I went out for a stroll in the morning. A young boy of about 10 or 12 years on a bicycle was distributing news papers to every houses around in that unearthly hour of the morning. It was normally too early for ordinary people to get-up from bed. He dropped a copy of the newspaper at every gate. He wore a helmet on his head. In Japan it is mandatory even for a cyclist to wear helmet. I was quite impressed and lovingly watched the boy doing his duty. How well-mannered are the Japanese; the boy bowed to me, a complete stranger, and greeted me saying OHAYO GOJAIMASH (our version of Good Morning). I reciprocated and responded OHAYO GOJAIMASH and the boy passed-by and continued his duty. For a while, I went to see the surrounding areas - it was all hills. It was evident that the hill sides were recently cut for widening the road. The portion of the hill sides, newly-cut, were plastered with cement to protect it from future erosion. Some portions not yet plastered by cement were covered by broad polythene-sheets as a precaution to unforseen rain.

At 9 in the morning we prepared our own food and had an early brunch preparing for our return to Mithila Museum. One notable social custom in Japan which is worth emulating by all of us is cleanliness. You cannot leave the house before it is properly cleaned. The entire plank-floor of the rooms had to be swept and throughly scrubbed with clothes. Similarly the kitchen had to be cleaned and all the utensils and crockeries are to be kept in their proper places. Last, but not the least, the bathroom and the toilet had to be left spick and span. In Japan there is nobody to perform such menial jobs, as we have in India. You have to do it yourself. It is a tradition in Japan that one has to work with one's own hands.

On our return journey, there was a heavy traffic jam on the way and we were late when we reached Mithila. Our friend Hiro-san (the man who was driving the Bus) began dozing while driving the Bus, probably because of the strain of driving almost 12 hours non-stop. I was alarmed lest an accident might happen.

We stopped for sometime in a Parking Area to have a cup of coffee. Since it was a long drive, our micro-bus required re-fuelling on the way. The gas-stations in

Japan were very interesting to observe and was quite a new experience for us. They were manned by young boys and girls in their teens. All the girls were dressed in short pant, like the boys. As soon as a vehicle draws-up for refuelling, one of them would jump up to the vehicle, sprinkle water on the windscreen and clean the glass. Another took the key from the driver and opened the fuel tank while another filled up the tank with gasoline. The driver had absolutely nothing to do; he did not even get down from his driving seat. Within five minutes, the entire exercise was completed by four or five smart youngsters. The key was then returned to the driver sitting in his seat with a bow and a smile on their faces. I wonder whether there is any other country like this - any nation as well-behaved and courteous as the Japanese.

It was 11 p.m. when we reached Mithila and had our dinner. The long drive made everyone feel tired and we went to sleep. The next two days we had no programme and we had to take a forced-rest at Mithila. That night there was a heavy downpour. Being a hilly country, rains are most unpredictable in Japan. Hardly had the scantiest cloud been sighted in the sky when suddenly it would rain. The next day the sky cleared and the surrounding hills of Mithila were bathed in sunlight.

That day I felt slightly unwell and told Hasegawa-san my wish to see a doctor. It was the morning of August 17. He arranged my visit to the doctor and I was to be accompanied by an officer of the Mithila Museum (Kiuseke-san). We had to cover about 30 kms upto Tokamachi City. My visit to the doctor brought to my mind the well-known joke of Khutta Macha (a popular comedian of Manipuri opera). It was a case of communication gap. What I spoke the doctor did not understand; what the doctor said I did not understand. He did not know English and I did not know Japanese. My friend who accompanied me was also equally uncomfortable with English. When I explained the details of my complaint the good old doctor listened attentively and did not speak a single word - he merely smiled throughout my narration, Finally, he gave me 10 tablets; God knows what it was. Speaking frankly, he did not at all know about my ailment. Needless to say, the Japanese tablets hardly contributed to ameliorate my sufferings!

On August 18, our performance was scheduled at Tokamachi City. By 3 p.m. Hasegawa-san took our entire group of 19 artistes to the Foreigners Registration office. Since we had to stay for at least 3 months in Japan, a special pass was needed. By 5 p.m. all formalities were completed and we left straight for the venue of our performance. As usual, many stalls selling Indian goods were set up. One small stall belonged to Hasegawa-san and another one to Mamoru Miura-san (the ex-judoka). Majority of the people who came to witness our performance were women and they made lots of purchase from the stalls before the start of our performance and during the 15 minutes intervals. Miura-san confided to me that sale for about 2 hours fetched him nearly 2 million yen (equivalent of our Rupees 3,30,000.00). Our performance in this city was organised by Hasegawa-san's friend, Higuchi san. The spectators were men and women of very advanced age (of the age of Govindji Temple - going devotees of Manipur). The auditorium had a capacity of only 400 people but the house was packed to capacity. It had a plank floor on which folding-chairs were placed. Everyone bought a ticket; there was no free pass or complimentary tickets, as we have in India. Everyone was expected to pay. At 10 p.m. we had finished our performance and returned to Mithila. The two terracotta artistes - brothers - Jagdish and Lala Pandit, had already prepared our dinner of rice, dal and vegetables. After taking it everyone retired for the night.

The next morning Hasegawa-san informed us that we would be going on an extended tour continuously for 43 days and perform all over Japan. During these 43 days we would not be returning to Mithila Museum. He advised us to take a light luggage of only the essential items of clothing since we had to travel for quite a length of time. He issued a big nylon hand bag to each artistes to put in their minimum essential things. He also issued one sleeping bag to every artiste, just in case we did not get proper accommodations and bedding in some places. This reminded me of our N.C.C. (National Cadet Corps) cadets going for an N.C.C. Camp. And like N.C.C. Cadets, we were also prepared for the long tour ahead. True to our fears, we actually gave a performance in a small Island having a population of hardly 200 souls. The accommodation in that island was inadequate for all of us and we had to sleep on a plank floor with our sleeping-bags on. They were also unable to provide bed-rolls for a big party such as ours. Our sleeping-bags came handy. In Japan, the normal form of greeting when you meet someone is OHIO GOJAIMASH - used only in the morning. It is equivalent of Good-morning in English. However, there is no equivalent of other forms of greeting such as Good-afternoon, Good-evening, Good-day etc. For greeting at other times, they use one general form of greeting - 'KONICHWA', roughly similar to 'Hello' in English. While meeting at night, the common form of greeting is 'KOMBANG WA'. It is not customary in Japan to shake hands on meeting. Instead, they exchange call-cards and bow their heads, bending their bodies at the waist. The English etiquette of introducing as 'My name is so and so' is unknown in Japan; exchanging the call-cards had done the job. A tip for people visiting Japan: take lots of call-cards which should display your name, address, telephone and fax numbers and e-mail address.

11

AKAKURA VILLAGE

August 20 was a Sunday and we were asked to give our performance at the Elementary School of nearby Akakura village at 3 p.m. The said village was about 15 minutes drive from Oike village where we had been staying. Hasegawa-san's 7 year old son (Munna) studied in that school and the show was arranged for the benefit of the students and teachers.

On arrival at the village we were greeted by a welcome-gate on which was written the word 'LENGSHIN BIRAKSHI' in Manipuri script. The organisers had managed to obtain the script from our boys without my knowledge. Akadomari is a tiny village and there was no proper place for our perfor-mance. It was therefore arranged in an open ground covered by a large carpet. Since the village was located in a cup-shaped small valley surrounded by lofty hills,

Welcome of Manipuri artistes at Akakura Village. Welcome banner in the background in Meetei script.

the sun was exceptionally harsh at 3 p.m. This caused much discomfort to our artistes. After we performed for about one hour, the Japanese also presented two items of their dance, in one of which mask was used. When the performance was over the village arranged a reception for our group at 7 p.m. It started by serving drinks profusely to the participants.

There was a mild surprise that night for us. The food offered to us was pure vegetarian while our Japanese hosts took sumptuous non-vegetarian food. There were mounds of fried chicken, Ham and Bacon heaped on their tables while we were treated with Cabbage, Broccoli, Spinach and Potatoes. Our artistes took the uninspiring dinner without much enthusiasm. It was strange that the guests should be treated so shabbily. I thought I would never be happy unless I knew the reason behind their unusual and apparent discrimination on the dinner table. I amassed courage and enquired the organisers. They replied in all humility and courtesy that it was done in deference to our religion - no disrespect intended. They were given to understand that we were Hindus and did not touch either meat or fish or drink and ate only vegetarian food. I did not quite know who had furnished this gem of an information to the organisers! It is only half true. Whereas our Vaishnavaite belief forbids us from taking meat,

Manipuri Drummer Boinao with a Japanese masked dancer at Akakura.

practically every young people of today freely consume non-vegetarian food and beer. Be that as it may, we had no way out of this predicament on that night. So while our Japanese hosts ate delicious fried chicken, Ham and Bacon, we had to be content with cabbages and salad to keep our religious flag flying in Japan! I had a strong feeling of suspicion, the needle of which pointed to Hasegawa-san. Even after the reception and dinner that followed were over, they refused to call it a day. Soon a social gathering was on its way - some persons danced and others sang while the Manipuris performed Thabal Chongba (a popular folk-dance of Manipur) in which the Japanese too, joined and made the day a truly enjoyable evening. The finale of the function was the playing of crackers. A variety of crackers were brought and all the artistes also joined our Japanese hosts in lighting and bursting the crackers in an outburst of joy and happiness. It was around 10 p.m. when we returned to Mithila.

12

TAJIMI

The next day there was heavy rain accompanied by thunder and lightning. Having nothing worthwhile to do, we slept till 5 p.m., but when we woke-up the sky was clear without any trace of cloud. A number of trees were growing in and around our house and I was able to identify some of them. There stood a walnut tree and also another tree, which in Manipuri, we call 'Wang'. The walnut tree was fruit-laden. I picked a few fruits and showed it to our boys. Around our house 'TONINGKOK' grew wild in abundance.

The weather in Japan is extremely unpredictable. A clear sky instantly becomes cloudy and turn into a drizzle. Since it had been raining throughout the previous night, the weather became unusually cold on that day. We took an early meal and by 10 a.m. left Mithila for Tajimi in Gifu prefecture, a distance of about 600 kms. It was 5 p.m. when we reached Tajimi. That day, both our drivers of the micro-bus and pick-up van were equipped with hand-held walkie-talkies. Both of

---

1. Toningkok–an aromatic herb, used as an additive to *IROMBA*, the most Popular Cuisine of Manipur.

them spoke to each other while driving and communicated about the direction of roads to be taken. Our micro-bus ran at a steady 120 kms per hour. We halted for some time in the Parking Areas and had tea prepared by ourselves.

On arrival, two of our female artistes reported sick. One had boils on her head and lip and the other had colic pain in the stomach. One drummer, who was nursing a fractured leg, had not fully recovered and was unable to participate in our programme. So, there were three liabilities on my head at the moment. I could do nothing except pray to God. The place where we stayed at Tajimi was in a hilly area. The organisers had already kept fish, knowing our weakness for the same. Among our artistes there was a Brahmin, who was an expert cook. He prepared the fish Manipuri-style, with the professional touch of a Brahmin. Good quality fishes were available in plenty, no doubt, but in the absence of mustard-oil, we had to cook it in soya-oil. Obviously, the taste was never the same. The next day, we went to call on the Mayor of Tajimi City (Mr. Masaya Nisidera). After meeting him we returned to our lodge and divided ourselves into two groups - one of female artistes and the other composed of male artistes. The female group was engaged in the workshop and demonstration of cooking (of Indian style food). The women of Tajimi were keen to learn how Indian foods were cooked from our girls. Not any specialised dishes but only the simple cooking methods of Dal, Subzi, fish and chicken were demonstrated. Not that they did not know the cooking of fish or meat, but their style and method of cooking them were quite different from that of ours. Our style of cooking, using Indian spices with a liberal sprinkling of chilly-powder, was quite alien to them; hence it turned out to be an entirely new-experience. The Japanese women also joined our girls in the cooking-process - some watched attentively the application of the Masala (spices) while some made queries and jotted down the points in their note-books. After this joint-exercise inside the kitchen, we took dinner together with the Japanese women. Some amomg them even attempted to eat with their fingers imitating us but failed miserably. It looked very much like little children learning to eat; all the food-stuffs falling from their mouth! It was quite an amusing sight.

The other group of boys conducted a workshop on our martial art, the Thangta, with the school children of the locality. The objective of the said workshop was the introduction of the Manipuri dancers to the children and to interact with them.

It was a sort of educational training for them to meet, talk and learn a slice of our culture during the short priod of the workshop. The children were taught how to play our Dholak and the elementary steps of our martial arts. Some of the earnest and enthusiastic children picked-up a few basic steps and displayed it. How pleased were the parents and the organisers! Grooming the little children as the future pillars of Japanese society and preparing them to become good citizens was one of the biggest agenda in the education policy of Japan. It was amazing to see how much the Japanese people had done for their children who were the future Pillars of the country. Just for a few hours' interaction and exposure to a group of people coming from an alien country and culture, so much amount of money had been spent. It was a workshop only in name; actually it was an opportunity provided to their children to meet us, talk to us, shake hands and ask any question which came to their minds. And how sweet and inquisitive they were and how well-read and well-informed some of them were! Such sights instantly brought to mind the contrasting fate of our own children. How different is our treatment and approach to child development. Children everywhere are very inquisitive. Wherever any foreigner or a high dignitary visits Manipur; the children are chased away; and are simply not allowed to come closer. The usual strain is "Drive the children out". Many foreign tourists visit Manipur, among them the Japanese are the most frequent group. But none of the small children of Manipur were given any oppprtunity to approach near the Japanese-not to speak of talking with them. Apparently, this was a great surprise to the Japanese visitors themselves. As they could not see any small children around in Manipur, one of the visiting groups expressed their desire to visit a children's Elementary school and they had actually distributed small presents to the pupils.

Our arrival at Tajimi coincided with an exhibition of the 50th year of World War II. Many ex-soldiers of the Japanese Imperial Army who had fought the Allies on the Burmese front belonged to Tajimi city. After going round the exhibition, I came to know that some of them had actually fought in the Kohima and Imphal front of the Battle. I expressed my desire to meet those who fought at Imphal sector. Immediately the organisers made telephone calls and made arrangements for me to meet the war-veterans. They also arranged one competent lady English interpreter for me.

The elderly soldier who had fought in Imphal was Shigetake Ando, now 78 years of age. He lived in Misaka district in Tajimi City. With the help of the interpreter he narrated his experiences. He was barely 22 years old when he joined the Imperial Army of Japan as a soldier. During World War II, they crossed the Kendwin River (in the then Burma) and reached the Indo-Burma border. By that time the Japanese Army was in a precarious condition. They had rations left for only 20 days. There was also a shortage of ammunition. In fact, they had only six artillery shells. In the end, they had nothing to eat; so much so that they had to catch snakes from the jungle and survive on it. On the other hand, the Britishers and the Americans (of the Allied Force) had put relentless pressure towads them. Not only with arms and weapons, they had started psychological warfare. Through loudspeakers and microphones, they persuaded the Japanese to drop arms and surrender to them. But the Japanese refused and fought till their last man.

Ando-san was now running a tyre-factory. The Japanese government pays him an annual pension of 7 lacs yen (equivalent of our Rupees two and half lacs) for services rendered to the Japanese Imprial Army. He seemed to be genuinely pleased to see us coming from Imphal, as if he was meeting his long-lost children. He invited us to visit his factory but we were, unable to oblige him due to our pre-occupation. But before he left, he gave us ten thousand yen as present. In Japan, paying cash as presents, in lieu of articles is quite common and customary. Wherever we went, we were loaded with presents; it is their custom to give presents to their loved ones. Everyone seemed to be keen to present something. In some places, people would approach us and asked if we would mind giving some presents. What a beautiful and loving national trait! It once again brought to my mind the difference of attitude in Manipur, may in India.

While the Japanese were keen to give, we in Manipur are eager to take. I feel disgusted when I see our men soliciting gifts from Japanese tourists visiting Manipur.

On that day we ran into some Indians also. They came to watch our workshop. One such person was Sri Harinder Pathak of Delhi who had married a Japanese girl. He opened an Indian Restaurant at Tajimi where Tandoori-roti, Tandoori-chicken, Chat, etc, were available. The Japanese people relished his Indian food. After a few moments, another Indian, whose name was Sri. Bharadwaz Sharma, also came to see us. He too, came from Delhi and was into the Hotel business in Japan. He was accompanied by a Japanese lady or rather, he was in attendance to that lady whom people called reverently as Madam Nakao. She seemed to be a wealthy lady and belonged to the elite of Japanese society. Her interest was in collection of fine arts and painting and she owned a private gallery. She showed a keen interest and affection towards the Manipuris and invited all of us to have dinner in the Tandoori Restaurant of Harinder Pathak. It was indeed, a very welcome invitation since we had been deprived of Indian food and delicacies for quite a long time. We accepted the invitation and thanked Madame Nakao for her hospitality.

13

OSHIMA ISLAND

We left Tajimi City the next day, i.e. August 24, early at 7 a.m. and carried packed lunch with us. After a drive of nearly 8 hours, we arrived at Imabari City in Ehime Prefecture at 5 p.m. En route, we crossed the longest bridge in Japan; I do not recollect its name. Our concept of a bridge is confined to a structure over a river. But the one we crossed was not an ordinary bridge; it connected two islands over a vast expanse of sea. At 5:30 p.m. we boarded a ship at Hashihama port of Imabari and after about 15 minutes reached a sparsely populated fishing island called OSHIMA having only about 20 families. Ms. Orita san and her friend Ms. Toyoshima san welcomed us. On that day a children's camp was being organised in that island by a voluntary organisation called 'Welcome Baby' working in the field of child development. Ms. Orita san was its Secretary General.

That night we, the 19 artistes from Manipur, I stayed on that island in the company of the children-campers. The vision of that night left an indelible impression on me. The distant sound of the waves beating the sea-beach could be heard. The powerful searchlights of the ships anchored at a distance, the intermittent hooting of their sirens and the clear night sky, was studded with a bright moon and countless stars. The bustle and the laughter of the little children at play resounded the tiny island and made that night a most memorable experience of life. I ran into two or three foreign tourists in that isolated and god - forshaken island - two youths from Canada and a girl from France. They too, joined in the fun and frolick of the children. I simply failed to imagine how and why they had come to that island which possessed practically nothing of tourist interest. I presumed that they were tired of city life and came there to have a taste of the calm and serene environment of OSHIMA ISLAND.

The way in which we spent one night at OSHIMA was itself quite unusual and a memorable experience. All the campers including our Manipuri group, stayed in small tents laid on a vast ground. No mattreses or bed-rolls were provided; everybody slept on the bare ground over which a thin plastic sheet was spread. The sleeping-bags which we took came handy at night. That was quite a new experience. Sleeping in the open with nothing over our heads except the starry night-sky as our roof created a strange feeling-of oneness and harmony with nature for which I cannot find any expression in words. It again reminded me of our NCC (National Cadet Corp) camps.

The sole purpose of bringing our group to Oshima Island was to have a close inter-action with the camping children. We did not present any regular cultural programme but conducted only a workshop with the little children. They were taught the elementary steps of Thabal Chongba which they picked-up without much difficulty. When we started the Thabal Chongba dance

(the most popular Manipuri folk dance form), our girls pulled eveyone by their hands and made them join the dance. The entire camp reverberated with the sound of drums and Dholok of the folk dance. The two Canadian youths and the French girl also joined. It was a carnival atmosphere where some people sang while some danced. That night provided me an opportunity to discuss with Hasegawn san and know him more intimately about his life and work.

Hasegawn-san narrated to me that 20 years ago when he was only 24 years, he left his comfortable hearth and home in Tokyo and came to live at Oike village. He was a person totally committed to the protection of nature and environment. Though he had his house at the posh Ginza street of Tokyo, he rented it out and settled in this village where the natural environment remained in its pristine purity undisturbed and unspoilt by human beings.

His one passion was to watch the moon every night. He was simply crazy of the moon and moonlit nights. That was some sort of a madness also, isn't it ? I mused, and this very musing was equally crazy, for moon means LUNA and the person under the influence of Luna means Lunatic! Suddenly this English word Popped-up in my mind at that moment, God knows, from where this stray thought sneaked into my brain! Well! Let us leave that aside for a while. Amidst the high-rise concrete structures of Tokyo, where could Hasegawn san watch the moon? Precisely for this reason, he left Tokyo and came to Tokamachi City (where the village Oike is situated).

After he settled in Oike, one day a friend showed him a Mithila painting by the renowed artiste, Smt

Ganga Devi of Madhubani in Bihar (India). That painting etched a permanent impression on his mind and fascinated him beyond any words of expression. That painting of Ganga Devi was the fountain of his inspiration behind the establishment of the Mithila Museum in Oike. The same painting still occupied a pride of place among the exhibits in the Mithila Museum. From time to time he organised exhibitions of the Mithila paintings and terracotta works to create an awareness of Indian Culture among the people of Japan. In sort, we can rightly recognise him as a genuine admirer and active promoter of Indian Culture in Japan. His ceaseless efforts of organising Exhibitions of Madhubani paintings and terracotta sculptures provided an opportunity to the people of Japan to have a peep into the rich cultural heritage of India in the shape of her paintings, arts and crafts of rural India and people of Japan began to appreciate and like them, too.

14

IMABARI

After spending one night at Oshima Island, we set sail the following day (August 26) at 10 a.m. after having an early brunch for Imabari. There the arrangement for the stay of all the male-artistes were made in the house of Ms. Orita san and that of the female artistes in the house of one Miss Toyoshima. During our long sojourn in Japan we had stayed in Hotels only on a couple of occasions, for reasons already mentioned in the foregoing pages. The organisers simply could not afford the luxury of a five-star Hotel everywhere.We ungrudgingly stayed wherever the organisers put us up. At Imabari we had a successful performance and quite a large crowd came to witness it. One of our boys of Thangta (martial arts) sustained injuries during the sword-fight and had to be taken to a nearby hospital for first-aid and for stitching his wounds.

Imabari

The next day we left Imabari after breakfast and reached Misaki port at 3 p.m. just at the nick of time when our ship was to pull its anchor. We had no time even to alight from our Bus and off-load our big costume-boxes. Hence, it was driven straight into the cargo-hold of the ship.Within a few minutes, the ship started moving. On that day we had to forgo lunch due to certain unavoidable circumstances. We carried some packed noodles in our bus and that served as our lunch for the day. Everyone felt tired due to the nature of the journey. It took only one hour to cover a patch of sea when we landed on the Kiushiu Island. Japan is composed of 4 big islands viz, Hokaido, Honshu, Shikoku and Kiushiu and many more small islands. From the landing point a continuous drive in our bus took us to Oguni City at 8 p.m.

Tearful farewell by Miss Orita san at Imabari

15

OGUNI

Oguni in Kumamoto Prefecture is a small town far from the madding crowd. The City Officer had already made arrangements for our stay. They had kept fresh fish, pork and beef for us. I politely returned the beef saying that beef-eating is a religious taboo for us. Oguni is surrounded on all sides by hills. The inhabitants of this place were farmers and agriculturalists cultivating large vegetables gardens. The town is famous for its wood-products.Their massive Town Hall was constructed entirely of wood, without any brick, mortar or stones. Nearby there was a volcanic hot-water spring where facilities were provided for a public-bath. There were many hot-water springs in Japan and plenty of people crowd there throughout the day for taking a dip in the hot-water for its therapeutic effects or simply for the pleasure of a hot bath.The hot water from the spring had been tapped and emptied into a nearby

Wooden Town Hall at Oguni

cemented tank. One felt fresh and invigorated afer bieng immersed in the hot water. The volcanic hot-springs are believed to possess healing qualities for treating certain disabilities. But there is a strict code of conduct. As already mentioned in the forgoing chapter, you have to step into these hot water tanks without a stitch on.

This town did not have many persons who could speak English and with whom we could communicate in English. They therefore, had engaged two English - speaking persons to help us in cooking of our meals. One such person was Miyazaki-san from Oguni itself who ran his own Indian Restaurant called 'Calcutta Restaurant' in Kumamoto City. He was a man who spent eight years in Nepal taking photo-copies of Buddhist scriptures written in Sanskrit and Pali. The influence of his long stay in Nepal was quite evident. He spoke Hindi and even his daughter was named Manju - a very popular and common name among Nepalese girls. He was accompanied by a youngman named Taro Yamamoto, to assist him in the cooking. Both of them spoke English fairly well. Yamamoto san's mother was a Doctor and his father a Professor in the Kumamoto University. It was quite a surprise for me that he could speak English as it was, so unusual among the Japanese. He narated to me that at the age of 10, he stayed for a year in Sweden where his father worked. He further told me that he was studying painting and fine arts and hoped to join the Okinawa Arts University for his higher education in fine arts. Both of them prepared Dal, Subzi (vegetables) and fish for our dinner.

Being a small town with a scanty population, there was no suitable auditorium for our cultural performance and we had to perform in an open lawn. Every spectator was seated on the covered ground after removing their footwear. When the performance ended, each and everyone joined in removing the electric light, P.A. system and the carpets and putting them in a small pick-up van. Then followed the cleaning of the entire ground for pieces of paper and other rubbish.

At the end of each performance on all the venues, 19 of our artistes including this writer as Manager of the troupe, would appear on the stage for a curtain-call. We bowed our heads in unison to the audience in the Manipuri style. As leader of the group. I was often asked to address the audience for about 5 to 10 minutes. I spoke, through an interpreter, about the history of Manipur and the origin of the Meiteis, their culture and traditions highlighting the close affinity between the Meiteis and the Japanese in respect of their religion, food-habits and even the outward facial features etc. The response of the audience was tremendous and overwhelming; loud clappings and cheers reverberated the place.

After our performance, we presented a modest Khamba-Thoibi doll to the organisers. We took a large number of such traditional dolls from Manipur with us. Wherever we performed in Japan, we would give these small presents of traditional dolls depicting Ras-Dance, Pung-Cholom, Manipuri Nat Sankirtam etc. The next day, after an early meal at 10 a.m., we left Oguni for Fukuoka where we would be participating in the International Ethnic Festival representing India. En route to Fukuoka, we spent some time to see a living volcanoe by the name of Mt. Aso. Japan abounds in volcanoes and many amazing sites formed by volcanic rocks. The awesome spectacle of a living volcanoe for the first time in my life left me spell-bound.

16

MT. ASO VOLCANOE

From early childhood, I had heard and read about volcanoes. In later life, I have seen many volcanoes in cinema and TV But I never had any opportunity to watch and be face to face with an active volcano from a close range. The volcanoe at Mt. Aso was more than a thousand years old and it was still burning. It was the volcanoe of my dreams and only destiny had brought me face to face with so spectacular a sight. The vast area around the volcanoe was full of steep rocks formed by solid lavas containing rocks and minerals erupting from it in the course of thousands of years. The area covered by the solid lava would approximately be 500 sq. miles. It was a vast expanse bereft of any tree or vegetations but covered by lush green grass creating the impression of a large green carpet. In the grassy meadows, a large number of cattle were let loose for grazing. All the cattle bore an identity mark on their legs to enable their owners to recognise their herd. The scene was awe-inspiring. For the conveneince of tourists, horses were kept ready which could be rented. Due to shortage of time, we did not venture to ride the horses and see around. Instead, we lost no time and ascended the mountain on foot spurred by the desire to watch the fuming and flaming volcanoe. As we went up, we could see a number of low sheds made of thick cement concrete. Those low huts looked like Military bunkers. In case of a sudden eruption of the volcanoe, these bunker–type low structures would serve as temporary shelters to the unwary tourists. The path leading to the peak of Mt. Aso was made of rocks formed by solid lava. There were many small stalls and wayside kiosks selling souvenirs and other curio items which were popular with the tourists.

Mt. Aso was found to be not a very high hill; we reached its top within about 15 minutes. It was from the peak that fire and smoke emaneted. Looking down from the mouth of the burning volcanoe into its abyss one was simply awe-struck at the sight of the giant flame and smoke leaping from the bosom of the volcanoe. The area around was surcharged with the strong stench of phosphorus and sulphur. Words fail me to describe the spectacle; it was fantastic to say the least. Every tourist appeared busy - some taking photographs and some making purchases of souvenirs. We also took a few snapshots and spent about 30 minutes on the peak of the volcanoe as we had to rush back to our bus; we were to reach Fukuoka City that very night.

17

FUKUOKA - INTERNATIONAL ETHNIC FESTIVAL

We arrived at Fukuoka at 4 in the evening. Arrangements for our stay had been made at SHIRO PUBLIC HALL where we slept on the plank floor with our sleeping bags on.

No sooner had we arrived, some of our girls started cooking while some joined the rehearsal since our performance was scheduled on August 31. The Manipuri group had been alloted 40 minutes for which we did rehearse some Manipuri songs like 'KOLOY NANGGI MANIPUR' or a Hindi patriotic song 'SAARE JAHAN SE ACHHA HINDUSTAN HAMARA ...'. The International Ethnic Festival was opened on August 29, at the 1000-seat capacity "SUN PALACE" Hall. The inaugural item was a folk song from Korea, presented by Miss Park Eun Hee. The Korean troupe consisted of only three members - one male artist played the flute, another female artist played the drum and the third one was

Spanish Dancer Miss Regla Sotto

Miss Park Eun herself, who was an University student. She wore the flowing traditional robe of Korea which almost touched the ground and resembled a long frock. This girl had a wonderful voice and the audience appreciated her song with loud cheers and clappings.

This was followed by a group from Spain who danced while singing gypsy songs. This group consisted of only three artistes. While their renowned gypsy dancer Regla Sotto performed, another male artiste provided the back-up music. Their performance failed to impress the audeince. After the Spanish programme, it was the turn of the Manipuris. While other groups consisted of four to five artistes, ours was a huge contingent of 19 artistes. The Festival's objective was to provide an opportunity to small ethnic groups to showcase their distinctive culture. No wonder, programmes of certain groups had no entertainment value. But entertaining the people was not the aim and object of the Festival; the most important thing being participation itself. I felt extremely proud and happy to have this opportunity of show-casing the distinctive culture of Manipur in the International Ethnic Festival (which gave international recognition of the Meiteis as a distinct ethnic group) Initially, the time alloted to Manipur was 40 minutes only. But the organisres had requested us to perform for 50 minutes, apparently because the other groups failed to impress and satisfy the audience.We opened our programme with the performance of our traditional item 'Lai Ikouba' which was well-received by the spectators. One reason for the Japanese people showing keen interest in our folk dances was the close similarity between the cultures of the two communities.

One of the participating groups which was of particular attraction to us was 'Gran Moun Lele' from

Reunion Island. This is a small island near South Africa sandwiched between Mauritius and Madagascar. Although geographically the island is within the African continent, it is under the administrative control of France as her colony. All the artistes of this group bore dark complexion and looked like African Negroes but their language was French. The very name 'Gran Moun' is a French word, meaning 'Old Man'. And true to its name, the lead-singer of the group was an old man of 65 years called Lele. The group was named as 'Gran Moun Lele' after its leader. Lele was not his real name; it was Philias Julien Earnest. He was a very popular and renowned artiste in his island-home. Although Re-union Island is a French colony, their dances and songs reflected a profound influence of African culture.

It was breath-taking to watch their powerful African dance on that night. The thunderous sound from the beat of the African drums, the sheer strength and stamina of their dancers, the like of which we had never seen before. This group consisted of four girls (they also sing while dancing) and eight drummers who played their drums with both hands. The thunderous sound of the eight drums by eight hefty young-men made our hearts almost come out. It was a sound not from the ordinary. It seemed to have emerged from the very depth of the earth and shook the huge hall by its vibration. I could not help admiring the strength and stamina of the four girls who danced their Hula-Hoopla for 40 minutes non-stop.

We had the pleasure of meeting them after their performance and took some photographs along with our artistes. The four dancers from Re-union Islands were of short stature, dark-complexioned and had wooly hair like the Negroes. We felt an intense urge to meet them, look at them from a close range and talk to them. The reason was that we felt it incredible as to how they could preform such a vigorous dance for 40 minutes without taking a single moment's break. I had no answer as to why we were emotionally so much drawn to these artistes from Re-union Islands. With the help of an interpreter (because they didn't speak English; they spoke only French), I came to learn that their forefathers came from South India. Suddenly, I felt a surge of emotions in my heart, of love, affection and a sense of one-ness with them. In my illusion, I saw in them as being Indians. Their history as narrated by them, was that about 200 years ago, their forfathers came from India to this Island as labourers in the sugarcane plantations there. And true, the artistes looked like Tamilians dark skin, wooly-hair, snubbed-noses and a rounded face. I had no doubt in my mind that they had originated from South India. At the end of their performance, Lele folded his hands and pronouneed 'Mata'[1], 'Pita'[2], 'Gurudev'[3].

The finale of that night's show was an improvisation in which all the participating groups appeared on the stage and danced in unison. Gran Moun Lele's group provided the song, background music and orchestra while the Korean, Spanish and the Manipuri groups danced. It was never a properly choreographed dance number; rather each group danced as they liked in an specified corner of the spacious stage. And bravo! the Manipuri group started with Thabal Chongba which took the entire auditorium by storm. Not only the artistes on the stage, but also the spectators from the auditorium rushed on stage and joined our folk-dance, while others

---

1, 2, 3. These are Hindi words, meaning Mother, Father, Preceptor

provided with rhythmic clappings from their seats. It was a spontaneous out-pouring of sheer joy and happiness and the entire auditorium reverberated with the sound of the African drums to the steps of Thabal Chongba. Unexpectedly, it turned out to be a most enjoyable and befitting finale of the International Ethnic Festival at Fukuoka. We extended our stay at Fuluoka for a few days more and gave performances in the vicinity of the city. There were some Indians among the audience of which the one closest to us was Mahammad from Calcutta. He was an engineer working in the reputed Japanese Electronics Company, Panasonic. He had married a Japanese girl and settled at Fukuoka city. He had a two-year old son. Mohammad and his family became our fast friend and inseperable companion. This was not so much because he hailed from India, but also because his Japanese wife looked just like a Manipuri girl. This had strengthened our bond. How much they genuinely loved the Meiteis could be evidenced from the fact that Mohammad came to Manipur as a tourist in 1996 along with his Japanese wife and son and visited the homes of each and every artiste of our group.

On August 31, Fukuoka City gave a reception to the Indian residents of the city including the students who were studying there. Besides the 19 of us from Manipur, a few athletes from India who came there to participate in the Universiade (the International Sports Meet for all University Students) were also present. Since they were keen to taste the Indian style of cooking, the organisers requested our female artistes to help them in cooking fish and chicken. Our style of cooking with a variety of spices and chilly powder turned out to be quite a novelty with the Japanese. They liked the food so much so that by the time we approached the dining table, there was hardly anything left for us; everything had been consumed! As already mentioned in the previous chapters, the Japanese food is bland and tasteless - containing no chilly or salt. Hence, their first experience of tasting Indian curry happened to be itself an unique experience for them. In the Fukuoka Universiade, India was represented by two sportsmen only - all from Delhi. One was a student from Delhi's Sriram College of Commerce participating in swimming and the other from St. Stephen's College in High jump.

On September 1, after giving one performance for the participants of the Universiade, we left Fukuoka for Yamaguchi. Mohammad and his wife also came along with us upto Yamaguchi. They had nothing in particular there; they went just for the pleasure of the company of the Manipuris. Thank you, Mr. & Mrs. Mohammad.

18

YAMAGUCHI

We were received at a place called Kumamoto in Yamaguchi. The cheif organiser was Nakano san who had already visited many Indian Cities like Bombay (now Mumbai), Delhi, Agra, etc. Immediately on arrival, we were asked to conduct a workshop for the children and after that to have lunch with the little ones reading in class II or III. Our artistes were seated in different tables with the children, talking to them while eating. It was simply amazing how much the Japanese were doing for child development. The children were encouraged to shake hands with us, talk to us and ask any question they liked about us. This again reminded me of the fate of the children of Manipur who were literally driven out from such gatherings. Observing the plans and actions taken up in Japan for child development, I had no doubt in my mind that all of them would grow up one day as responsible citizens of the country in future.

On September 3, we gave our performance in an open-air theatre called 'Big Shell'. Indeed, the architecture of the theatre resembled a sea-shell. The attendance on that day was very poor. There were hardly 200 people, mostly young children. At Yamaguchi, arrangements for our stay was made at an expensive

Hotel which had its own facilities for hotwater bath. But the peculiar custom of the Japanese is that such a first class Hotel did not provide any cot or bed stead. One had to sleep on the plank floor covered by a Tatami (Japanese mat made of fine reed) on which bed-rolls and mattresses were spread.

In the morning I took a stroll of a few miles. I found paddy being planted near a beautiful building. The paddy fields were enclosed by a low wall of brick and mortar. It was time when the paddy in the fields were just ripening. I remembered of my own paddy fields in Manipur which must also have been ripened by now. The people of Yamaguchi seemed to take a very keen interest in kitchen gardening. *Paankhok, Yendem, Yenam Nakuppi, Naapakpi etc. were extensively cultivated. In Manipur, every family by and large, possess large plots of land for construction of their houses and kitchen-gardens. But in Japan, land is a scare commodity and it is impossible to own large plots of land due to its exhorbitant price. None of them possessed *Ingkhols* (homestead land) as big as ours. These houses were small but extremely neat and clean and well maintained. Invariably, I found flower-beds and flower-pots and a well-manicured pine tree in front of every house. And that speaks volumes about the high aesthetic sense of the people of Japan.

On the next day September 4, we left Yamaguchi in the afternoon. In course of our long drive, we passed through a number of tunnels whose length we had meticulously recorded. Some of them were as long as 6 kms. The interiors of these tunnels were lighted by

---

* Endemic vegetables of Manipur

powerful Halogen lamps and were fitted with huge exhaust fans to suck out the poisonous gas and fumes from inside. One could have an idea of the size of these tunnels from the fact that two vehicles could easily ply from either end. The engineering skills of the Japanese were simply marvellous. At 5 p.m. we reached Yamada town in Miyazaki Prefecture.

19

YAMADA

On arrival, I was told that our place of stay had been arranged at a Budhist temple called Shojo-ji. The priest of the temple Amako Gensho san had been waiting to receive us. The prospect of staying in a 'temple' had brought serious misgivings in my mind. The concept of a 'temple' in my country is a sacred place where the idol of the deity is being worshipped. None even dare to climb up the steps of the temple, let alone live inside it. It is a sacrilege to eat, drink, smoke and engage in revelry inside the precinct of the temple. Similarly our priests are holymen dedicated to the service of the Lord and their life style is different from ordinary folks like us. In this way an aura of sanctity and divinity envelops a temple. But in Japan the concept is wholly different from ours. They were not bound by the rigid and orthodox principles like

Teaching of Thabal Chongba to the children of Yamada Town

us either in their Budhist temples or the Shinto shrines. The temples in Japan had large and spacious complexes. Close by the sanctum sanctorum where the idol of Lord Budha was being installed, there were a number of guest rooms all around it. Similarly their priests were ordinary common people like us; they had their wives and children, did take meat and drink beer. They even carry a camera in hand and take photographs of us! (which is unthinkable in our country). This was another culture shock we had encountered in Japan. They were persons who only performed the routine job of a priest for which they were being paid a monthly salary, like any other government service in India.

By 6 p.m. the priest (Gensho san) came to the temple and sounded the temple bell, lighted candle and joss sticks in front of the grand idol of the Budha and began chanting prayers. We, too, joined the prayer and sang the devotional song - 'Ram-Dhun'. The Hindus also recognised Lord Budha as one of the gods of the Hindu pantheon; He is wordshipped as the tenth incarnation of Lord Vishnu. We spent 5 days and 4 nights at Shojo-ji temple and gave performances in and around the town. Amako Gensho- san happened to be a personal friend of Hasegawa-san and as such our accommodation was arranged at the Buddhist temple without much expense. It was because of such assistance from voluntary arganisations and well-wishers that made it possible for us to perform in more than 40 towns and cities in Japan.

20

KAGOSHIMA - MIYAKONOJO

Kagoshima City was about 2 hours drive from Yamada town. On September 6, Manipuri cultural programme was performed in the CHUO PUBLIC HALL of that city. In front of this large hall, there stood a massive stone statue of the legendary martyr Takamori Saigo who was born in that city.

As usual, a bazar selling Indian goods was set up at the entrance of the hall. Hasegawa-san, too, had brought a large quantity of Indian goods for sale in this place. Wherever we performed, one stall of Hasegawa san and another belonging to Miura Momura-san was a must. Those who came to watch our performance bought the Indian goods from these stalls just before the start of the performance and also during the interval. As stated already, the items sold there were nothing special; ordinary Indian Salwar-kamij, Bangles, Joss-sticks, Cheap necklace, Brassware etc.

The next day we performed in a small town called Miyakonojo inside a Buddhist temple. Everyone took their seats on the plank floor. Before entering the place, everybody removed their footwear. The volunteers, at the entrance provided a celophone carry-bag to each of them. They put their footwears into the bag and took it inside the hall. In Japan it is not customary to enter one's home or a temple with their shoes on. But in offices this restriction is relaxed. The spectators of that day appeared to be very knowledgeable and appreciative about our art and culture. During the performance, they too expressed their appreciation of the artistes from time to time by prolonged clapping.

On that day (September 7) Hasegawa-san took some of us, including a few female artistes to call-on the Mayor of the City. A very memorable incident, although a minor one, happened during that visit. In the office of the city Mayor, a water-melon was on display which weighed 40kgs. It was a pleasent surprise to see a water-melon weighing one maund. After completing our performances in and around Fukuoka City, the moment of our departure from Shojo-ji Temple arrived. The next day we had to continue our journey to fulfil our engagements in other cities. Amako Gensho san, the priest of the temple, gave us a farewell dinner. His wife Soko-san and his ten year old daughter Chihiro Amako presented small gifts to each and every artiste and bade us a loving farewell. We could never forget the love and care which the priest's family, his elderly father and mother showered on us during our short stay in the temple.

Dance Workshop with the children of Yamada Town

21

DRIVE FROM YAMADA TO AKO

Before leaving Yamada, Hasegawa san forwarned us. "Today you have to sit in the Bus for quite a long while; our destination is pretty far from here!". Our destination was Tatami city but since it would be difficult to cover the distance in one day, we had to break journey on the way in a small village called Ako in Himenji City. This drive from Yamada to Ako village took eleven hours; the distance must be about 800 kms. En route to Ako, we passed through the cities of Kobe and Hiroshima. As we neared Hiroshima , I was thrilled at the prospect of fulfilling my long cherished dream of visiting Hiroshoma. We spent long hours at Hiroshima (meaning world island) visiting Atom Bomb Museum, Atom Bomb Dome, Peace Park etc. A visit to Japan is not worth the trouble unless you include in your intinerary the city of Hiroshima. Tokyo, being the capital of Japan may be well-known and famous, but there is nothing much to see there. It is just like New Delhi. The famed Ginza street is not very diffrent from our own Connaught Place of New Delhi. Yes, the only diffrence I noticed was that there were no pedestrians on the road; everywhere there were cars and cars only. I shall be devoting a full chapter exclusively on Hiroshima in the subsequent pages. We crossed a number of tunnels while driving on the long stretch of road in the hilly terrian. The Highway passed through uninhabited areas dotted with large junk-yards where old and unserviceable items were deposited. One reason why they did so was that they could not waste their valuable time in making repairs. Besides, I was told, it was much more economical to buy a new car rather than repairing an old one. And it must be true because during our three month's extensive tour across the length and breath of Japan, we had never come across a workshop or garage for repairs of motor cars. This brought to my mind of an encounter in Imphal with a friend who was simply crazy about Japan. A few days before our departure for Japan, he requested me to bring a leather cover of his Minolta Camera from Japan. You can well imagine my predicament. In a country where almost new cars or Television sets, but slightly defective, are thrown in junk-yards, where should I search for a leather-cover of an old Camera ?

That did not necessarily mean that there are no second-hand shops in Japan. They do not use the phrase second-hand; instead, they use the more respectable term, 'USED'. Being an affluent people, as soon as a new model appeared in the market, be it a car or a Camera, they dispose off their old models in second-hand shops and buy the new models. Their used-cars are lined-up with a price-tag attached on them, in a vast field as big as our Pologround and are sold in bulk. Nikon is the most popular brand among Cameras in Japan. As soon as a new model is introduced in the market, many well-to-do people dispose off their old Cameras.

Driving from 7 a.m., we reached Ako in Himenji City at 5:30 p.m., barring about three hours for our visit to

Hiroshima. The arrangements for our stay at Ako was made at a newly-constructed, but still unoccupied house, belonging to a friend of Hasegawa san. Similar hospitalities were extended to us by Hasegawa's friends in many other cities also. The next day we prepared early lunch since our journey upto Atami City would take no less than twelve hours.

While our artistes were preparing lunch, I took a stroll in a nearby Park. Animal figures made of cement concrete were installed in every nook and crany of the Park. Among them, I found the figures of the three celebrated monkeys of Gandhiji - speak no evil, see no evil and hear no evil. After food, we left Ako village and drove along the wide Highway heading for Atami. The Highway passed through hills and the landscape around was quite similar to our Imphal-Dimapur road. The only difference was that their roads were far superior to ours; they were broader and had two separate lanes. The roads were as straight as an arrow; having no curves or bends on the way. Whenever a road struck a hill, it did not take a detour or bend around the corner. In other words, the roads did not follow the contours of the hill. On the contrary, two tunnels were just bored, wide enough for the vehicles to pass through it. The two tunnels ran pararell - one meant for on-going vehicles and the other for in-coming ones. As soon as the tunnel was crossed there was a sudden drop of 100-200 feet, forming a formidable gorge. Bridges were constructed to ford the deep gorge whose supporting pillars were not less than 200 feet high. The roads were planned to maintain a smooth gradient and facilitate safe driving. It was indeed a tribute to the engineering-skill of the Japanese in road-building. The long stretches of the Highways passed through uninhabited hill areas. For miles, no sign of a village was in sight. Some sections of the Highway needed traffic control and signal. That job was performed by robots.

Another experience we came across, in the course of our long drive, was the unending rows of flowers being planted on both sides of the road for miles and miles. How many of us maintain a flower garden in our own homes? But in Japan, not only in their homes but also even on roadsides, people planted flowers. This is unthinkable in India. It exemplied the high aesthetic sense and love of beauty of the people of Japan. This was quite a pleasant surprise to me and I enquired as to how was it possible to plant flowers for so many miles. The reply was even a greater surprise for me. These flowers were planted not by any government agencies but made it possible through people's participation as a voluntary venture.

On our way back from Ako village, we halted for sometime near the Shiga Lake, considered to be the largest one in Japan. In this place, excellent facilities for tea and refreshments were provided. We prepared our own tea inside the Bus. By 6 p.m. we reached Atami in Gifu prefecture. In this journey, we made it a point to take count of the number of tunnels we had passed through on that day. It turned out to be forty in all, some of them stretching upto a length of six to seven kilometers.

Founder of Johori, Meishu Sama, also called Mokaichi Okada

22

ATAMI

At Atami we stayed at a Hotel built near a sea-side. Our dinner had already been prepared by Mrs. Hasunama san who had reached there in advance. It was a simple fare of rice, dal and vegetables. Among the staff of the Mithila Museum Hasunama san only had knowledge of Indian style of cooking because of her long stay in the Madhubani District of Bihar, India. There were only a few people in Japan who were familiar with the Indian cooking method. One of the greatest problem for us was that we were never comfortable with the Japanese food and it was solved by our self-cooking. The long journey made everyone tired and we went to sleep.

The next day, September 10th, Manipuri cultural programme was performed at the M.O.A. Museum (Mokaichi Okada Association) of Atami City at 2 p.m. M.O.A. is a religious organisation The M.O.A. Arts Museum is the biggest Arts Museum in Japan which, among others took-up nature-farming and also art. People from all corners of the world would come here and worked as volunteers. One such volunteer was Miss Sarah Martins from Portugal, who presented me many literatures on M.O.A. She had been very nice and friendly especially with us. She informed me that her grandfather came from Goa of India.

While introducing M.O.A. to my readers, I would like to make a special mention of a programme named Johori Fellowship. The man who was responsible for introduction of Johori was Meisu Sama (his earlier name was Mokaichi Okada). In simple terms, Johori is a system by which the vital energy of one person is being transmitted to another person to infuse mental tranquility, peace and sound health. A Johori session took about 20 minutes: One needed no physical contact. The person who would transmit his vital energy is called Johori channel. He would raise his right hand and the invisible energy from his body would be transmitted from his palm to the person sitting opposite him an a stool. Apparently, it somewhat similar to hypnotism and mesmerism but in reality, it was not. It was a distinctive method of treatment. I watched a Johori session for sometime. There were streams of people coming there to attend this session. Each of them were taken to different rooms and the volunteers acted as Johori channels and treated the patients of their ailments. Here the use of the word 'patient' may not be quite appropriate since many of the clients were not sick people. Quite a large number of them came for the Johori session for the simple reason of obtaining peace of mind and eliminating mental anxieties etc. Not only in Japan,

Johori session

Johori fellowship had spread in many parts of the world because of its general acceptance by the people as a beneficial system.

As soon as our performance was over, we left Atami immediately by about 4 p.m., and headed for Tajimi City, a distance of about 5 hours drive. Mrs Oyanagi san, who was the organiser of our cultural programmes had been waiting for our group to arrive. She made an all out effort to make our stay at Tajini comfortable and made us forget the hardship of the long journey.

23

HIROSHIMA

Whoever has not heard of the City of Hiroshima and the saga of the Atom Bomb ? For every tourist visiting Japan, Hiroshima is the first priority in their intinerary. Quite obviously, we too were extremely keen to visit Hiroshima. But unluckily, the famous city was not included in the intinerary of our cultural performances. However, as already mentioned, we had to pass through that city while coming from Fukuoka to Ako village. Taking advantage of this chance of a life-time, we broke our journey at Hiroshima and visited the A-Bomb Museum and other sites associated with the bombing of the city. What a difference between my pre-conceived image of Hiroshima before I visited the city and the Hiroshima as it stands today. The city had been completely wiped off from the face of the earth on that fateful day of 1945, when the United State of

View of Hiroshima after the A-Bomb

America dropped the A-Bomb. For the first time in the history of mankind, by dropping the A-Bomb, the U.S.A. had completely destroyed the city. Thousands of people had lost their lives; all the houses and buildings were razed to the ground. That was the picture which had been permanently etched in my mind since 1945. But surprisingly today, like a Phoenix rising from the ashes, Hiroshima had been metamorphosed into a vibrant and ultra-modern city, with vast stretches of sprawling Parks of lush green vegetations and tall buildings. I really could not believe my own eyes. "Is this the same city which is said to have been doomed after the American bombing with an Atom Bomb"? I would ask myself again and again out of sheer disbelief and amazement. The ground zero (the actual spot where the A-Bomb was detonated in 1945) which was charred beyond recognition, had now been transformed into a lovely Public Park known as 'Peace Park'. Almost sixty years back from today, a powerful Atom Bomb was dropped on the city of Hiroshima which had completely obliterated it from the face of the Earth and left no trace of the once beautiful city. The only monument, and a grim reminder to that moment of history, is the damaged 'Atom Bomb Dome' which stands today as a mute witness to the horrors that had been perpetrated by the United Stated of America on the innocent people of Japan. This building (A-Bomb Dome)

Atom Bomb Dome

was originally the Hiroshima Prefecture Industrial Promotion Hall. For the benefit of the tourists visiting Japan, this mutilated structure is preserved exactly as it was seen after the bombing on August 6, 1945. It is being kept untouched and without repair. On that fateful day, the A-Bomb was detonated just 600 metres above that building resulting in its complete destruction; only the skeletal iron-frame of the dome remained. The Atom Bomb Dome had been declared as a World Heritage Site by the United Nations Organization.

The U.S.A. had already programmed the A-Bomb (they named it 'Little Boy') to detonate in the air before its impact on the ground for enhancing its devastating effect. The population of Hiroshima then was about 3 lacs 23 thousand. Within seconds, nearly 80 thousand people were killed and double the number were maimed. Out of about 76,000 buildings, 48,000 were completely razed to the ground. The destructive power of the Bomb was its extreme heat and flash. Within 1000 mts. of the hypocenter rocks, stones, iron, etc. in fact , everything melted. All those people killed in Hiroshima were not solely due to the heat and flash. People who were exposed to the Atomic radiation within a radius of 2 to 3 kms. , too, gradually succumbed to its deadly poison and met a slow and painful death. The horrendous experiences of the people of Hiroshima on that fateful day would be beyond description. How men and women, with their bodies severely burnt, ran half-crazed in search of water ; the thirsty souls would throw themselves into the nearby Ota River to quench their thirst only to find that the river water had become

'Little Boy' - the A-Bomb dropped over Hiroshima

radioactive and hence, unfit for human consumption. Many of the people who were exposed to the extreme nuclear heat and flash, and whose houses were all destroyed and burnt, died due to sheer thirst. Among the dead, there were a number of schoolchildren within the age group of 12-13 years. On the day of the bombing, school children had been drafted for demolition work as a precautionary fire-fighting measure against bombing.

In Japan, 8:15 AM is rush-hour when streets are crowded with office-goers. The Americans chose this crucial hour to drop the A-Bomb over the city of Hiroshima. I have already mentioned briefly about the intensity of the heat generated by the A-Bomb. It was

Devastated Hiroshima City after the explosion of Atom Bomb.

estimated that the temperature at the hypo-centre (the spot where the Bomb exploded) was between three thousand and four thousand degree Celsius. Within seconds the entire city of Hiroshima was engulfed by fire and transformed into a blazing inferno. Everything in sight was burnt-up by its incredible heat. The fire generated a chain reaction. Because of it, a strong wind swept across the city, followed moments later by radioactive black rains for about one or two hours which killed all the fishes and organisms in the rivers and lakes. There was not a drop of drinking water available and thousands died due to thirst. People were completey bewildered as to what catastrophe had befallen on them (it was much later that the Japanese people came to know that the bomb dropped by the U.S.A. on that day was an Atom Bomb).

Those who were comparatively far away from the hypo-centre, though saved from the extreme heat energy released by the atomic explosion, lost their hair within 2 to 3 weeks due to exposure to atomic radiation. An Atom-Bomb is not a conventional bomb as we understand today. It is precisely for this reason that people dread it the most. The destructive power of this bomb is beyond human imagination. I am here reminded of the observation of J. Oppenheimer, Director Manhattan Project (which undertook research on making the first Atom-Bomb). After sucessfully completing the Bomb, they conducted a test in a desert area. A tiny A-Bomb, the size of a baseball was placed on top of a 100 feet tall iron tower and detonated. On inspection after the explosion it was found that the entire area around the desert became a vast sheet of glass while the iron-tower disappeared as vapour! That would give an idea of the intensity of heat generated by an atomic explosion. At that instant, J. Oppenheimer was reminded of a passage from the Hindu Scripture, the Gita - "Now I have become death, the shatterer of worlds". (Ref. Peter Goodchild: J. Robert Oppenheimer: Shatterer of the worlds, Boston 1981 pp. 162). It may be a gross under-statement to call it a bomb; it is in fact the shatterer of the worlds incarnate. As a devout Hindu, I felt pride and pleasure for an entirely different and bizarre reason: of all the holy scriptures of the world, how did the sloka of the Gita (the 32nd Canto of Chapter 11 of the Gita named "Vishwarup Darshan" which runs as:- 'KALOSMI

LOKAKSHAYA KRIT PRABIDDHAH, LOKAN SAMAHAR TUMIHA PRAVRTTAH' which translates: (I am the world-destroying Time, grown in stature, and now engaged in annihilating the creatures.) popp up in the mind of J. Oppenheimer at that moment?.

Albert Einstein, the inventor of the celebrated formula $E = mc^2$, who is known as the father of Atom Bomb, on seeing the destructive power of the Bomb, regretted and remarked remorsefully- "If I am to mention one single mistake I have committed in life, it was the making of the Atom Bomb". Let us now analyse why the U.S.A. had taken recourse to the A-Bomb in Japan. In other words, was it absolutely necessary to drop the A-Bomb in Japan in August, 1945? To this day, the question is raised and the debate is still unending. By August, 1945 World War II was almost coming to an end. In fact, by May, 1945, the war was in its final stage. The Allied Power was fully aware that Japan no longer had the capacity to resist and it was only a matter of time when they would surrender. In the western front, Japan's ally - the Nazi Germany - had been decisively defeated and it was impossible to resist the combined onslaught of the Allies by the Japanese Imperial Army single handedly. That being the situation, a question arose as to the rationale which prompted the U.S.A. to use the Bomb over Hiroshima killing thousands of innocent citizens and completely devastating the City. There might be a number of considerations but one thing was quite obvious; the Americans, by their overt actions, revealed a hidden agenda to test the efficacy of the Bomb on human beings. Otherwise, how would they explain the reasons for selecting a civilian target - a thickly populated city during peak office hours, instead of either a naval base or a military installation or a war-factory where weapons were manufactured? The U.S.A. made the A-Bomb by spending an astronomical amount of public money for use in the war, and in case it ended without having an occasion to use it, it was feared that there would be a public hue and cry in the U.S.A. and lots of uneasy questions would be raised subsequently about the rationale of the Bomb. In fact, they were in a hurry to test the Bomb.

Since the nuclear holocaust happened almost 63 years ago, obviously, the present generation has not seen it. Yet for posterity, and humanity at large, to realise the horrors of an A-Bomb, collection of various items associated with the Hiroshima Atom Bomb, are now being displayed at the Peace Memorial Museum (the correct name is Hiroshima Peace Memorial Museum). I led the group of Manipuri artistes and entered the Museum.

Atom Bomb Museum

It was, indeed, a remarkable monument - a poignant reminder of the ghastly scene on the day of the bombing. Of all the exhibits therein, the one which struct me most and created a permanent imprint in my mind was a small wrist-watch, whose hands stopped at 8-15. That tiny mute object proclaimed loud and clear for the whole world to listen that at 15 minutes past 8 O' clock on the fateful morning of 6th August, 1945, America dropped the terrible A-Bomb over the City of Hiroshima and levelled it to the ground.

In a corner of the Museum, I set my eyes in a bunch of human hair, completely removed from the scalp by atomic radiation, still showing the same original shape. On first sight, it was mistaken for an artificial object like wigs or curls used in theatrical performances. A number of burnt uniforms of school children were also displayed. Human skins and nails which came-off from exposure to atomic radiation on that fateful day, were also among the exhibits. It was, a heart-rending spectacle. Huge iron girders and beams melted into grotesque shapes and sizes, bore a grim testinomy to the extreme heat generated by the monster Bomb. A certain non-descript person was sitting on that-fateful day, on a stone slab in front of a Bank awaiting for its opening. That very moment the A-Bomb exploded, the unknown person just vanished into thin air; only the spot where he was seated, could still be seen blackened. There were many more tale-tell exhibits inside. Indeed, whosoever had visited this Museum would never dream of using the A-Bomb.

Photography was strictly prohibited inside the museum. But there were no security personnels inside. Inspite of the warning against photographing the exhibits, and my admonitons to my boys to adhere to the rules, the lure of the awesome exhibits were too over-to wering for them. Some of the boys (all had a camera each) could not simply resist the temptation, and taking advantage of the absence of security guards, discreetly took some photographs of the exhibits therein. They were very keen to show the rare photographs to their near and dear ones at home. There was also no chance of our coming back to Hiroshima again. When we all came out after visiting the Museum, their faces were brimming with smile at the minor coup of successfully taking some clandestine photographs of the Atom Bomb Museum exhibits. They congratulated and patted each other on the back on their achievement, undetected by Japanese Security Guards. But their joy was short-lived. When the films were processed. Lo and behold! there was not a single picture in the frames; all the negatives turned out to be completely blank! I recall this small incident to highlight the height of Japanese technology in the field of electronic engineering. They too, were fully aware of the human weaknesses. Inspite of the warning, they still suspected that some of the visitors would surely attempt to photograph the exhibits. That is why they had secretly installed some electronic device to neutralise clandestine photography.

Peace is not an empty slogan but the very creed of the Japanese people. Their clamour for total abolition of Atomic weapons is not for the purpose of propaganda but the sincere cry of a wounded soul. The sight of the various Peace monuments in Hiroshima would convey the eternal message of peace to every visitor to the city. To mention the name of some of the monuments: Pond of Peace, Flame of Peace, Peace Memorial Museum, Peace Tower, Peace Bridge, Stone Lantern of Peace, Children's Peace Monument, Figure of the Merciful Goddess of Peace, Peace Park .... etc.

As already mentioned, the site where the Atom Bomb exploded has now been made into a beautiful Park and named as 'Peace Park'. Peace rallies and sit-in protests are held in this Park round the year. On the day of our visit to the Park, about 20 to 30 people with paper flags in their hands, held a Peace rally. It reminded me of the "Boat Club" of New Delhi, where one group or the other held sit-in protests throughout the year. Another notable

Hiroshima Peace City Memorial

feature of the Park was the presence of countless number of pigeons. In 1945, a few pigeons were released as symbol of Peace on the innauguration of the first Peace Festival. In course of the last 50 years, they have multiplied and the pigeon population which has now reached a staggering one lac and more. Tourists visiting the Park would buy feed from a nearly Kiosk and field the pigeons. Out of the numerous monuments raised inside the Hiroshima Peace Park, the most significant one is the Hiroshima Peace City Memorial. There is a stone coffin in the centre of the monument in which is kept a register containing the names of those killed by the Atom Bomb. The second most significant memorial is the Children's Peace Monument, also known as the Tower of the Paper Cranes. This menorial is the brain-child of a girl called Sadako Sasaki. On the day of the Atom Bomb, she was only two years old. After ten years she was affected by Leukemia. While lying in her sickbed in the Hospital, she made more than a thousand paper-cranes from the wrappers of her medicines out of a belief for a swift recovery. Her classmates too joined her in making thousands of paper cranes. But Sadako succumbed to the disease in 1955 caused by atomic radiation. The heart-broken class mates of Sadako decided to build a

Children's Peace Monument

monument in memory of their friend. Making paper-cranes is almost a national pastime of the Japanese; they call it Origami. A cement and concrete monument was raised, on which stood a bronze figure of a little girl, holding a giant origami in her hands.

Whoever visits Hiroshima should take a vow never to use the Atom Bomb again on human being. Unless a complete ban of the nuclear weapon is imposed, the fear of total annihilation of the human race would haunt us forever. Never forget that this is not a mere bomb; it is doom personified.

24

TAJIMI-2

At Tajimi we were lodged at Sannokura Public Hall. Plenty of sea fishes were kept ready for us (they knew our weakness for fish). The memory is still fresh in my mind of how the fish on that day was prepared in the traditional Bhandra style (specialised method of cooking fish on ceremonial occasions) by our Pung artiste, Manisana Sharma, and everyone enjoyed an elaborate dinner. Somehow, I did not relish it because it was cooked in Soya oil, as mustard oil was not available in Japan. No wonder, the taste was well, tasteless!

That day, since the early hours of the morning, the tranquility of the night and our deep slumber was disturbed by intermittant gun-shots. Coming as we were from Manipur, where shooting and killing every night had become common-place, we were naturally alarmed and anxious to find out what it was about. Every five minutes the sound of gun-fire resembling that of a double-barelled breach-loading gun, wás heard quite close-by. Getting up early in the morning, I took a stroll around the area with two of our boys to trace the source of the sound. O, it was nothing but a mechanical device an automatic gun, to scare away birds from the paddy fields. There was a paddy field close-by our place. Paddy was then just ripening. For protection against birds, a contraption was installed in the middle of the field. Intermittant gun-shot sounds emanated from this small machine after regular intervals for driving out birds.

The following day onwards, we were shifted to a large Public Hall named Chikumura Earth Village. We stayed for four days in this place and gave performances in and around the nearby towns and cities. A two hours drive from Tajimi City took us to Yokaichi where we ran into some Indians who came to witness our performance and we chatted with them. One pleasure of meeting Indians in Japan was the opportunity of conversing in our common language viz, Hindi and English. There were not many among the Japanese who understood what we spoke (in English); it became a major irritant for us. Hence we mostly communicated in sign-languages, supplemented by lots of giggles and laughter.

A remarkable aspect of life in Japan was that there were no petty thieves. I asked, out of curiosity, the reason for this. They replied - what was the use for thieving when everyone possessed everything they needed. I thought what he told me must be true: most of the houses which we had seen were not under lock and key, as we do in India. Most of the doors were push-in type; they were just being pulled in and left. Government Offices and Public Halls were there but no chowkidars or night watchmen were seen around. Employees of a Government Office, at the end of their days work, switch-off the electric lights and fans, shut the doors and windows (they did not wait for the chowkidar to do this job) before leaving for their homes in their own transport. The next day at exactly 10 a.m., they would arrive at the office, open the door and immediately sat down for work. All the costly gadgets and other valuable items inside the office were just left unguarded at night and they were never stolen. It was a pleasnat surprise for me to observe all these things. The very Public Hall where we stayed would, in India, require at least ten to twenty chowkidars/night-watchmen. But in Japan, there was no need for them and they considered it a sheer wastage of precious manpower. It was also an indication of the civic sense of the people in its highest form. I could not help comparing it with what was prevailing in India.People working in Government offices do not care to switch off even the lights and fans, not to dream of closing the doors and windows, before they leave their offices. Herein lies the difference. The difference is in the quality of the people. As for tall buildings and lovely gardens, India is not lagging behind Japan.

I am sure the reader must have been curioús to know whether there were other devices in the offices in Japan to detect an unauthorised entry into the offices at night. Although it bears repetition, the readers would pardon me on harping on the excellence of the Japanese in electronics. The large Public Hall, inspite of the absence of any night watchman, was fitted with an electronic device to warn the fire-brigade and the Central Public Control Room in case anybody sneaked into the office after 10 p.m.

On September 9, Manipuri Thang-Ta and Dhol-Dholak was performed at Tajimi City. In the foyer of the Hall as usual, there were a number of stalls selling cheap Indian wares like Bangles, Bindi, idols of Hindu Gods and Indian clothing like Salwar kurta etc. The stall of Momura Miura san was always there. He and his stall

Minoaki Porcelain

of Indian goods followed us as an inseparable shadow. At Tajimi the number of such stalls were quite many. Harinder Pathak, also opened his eatery and restaurant. His Japanese wife, drapped in a green silk Sari, assisted him. Another Delhite (his name was Adarsh), who had his restaurant at Nagoya, also came to Tajimi on that day to open a Restaurant. In Japan, 100 or 200 Kms is not considered a significant distance. He had engaged two Sari-clad Japanese girls as sales-girl. But by far, the most interesting and remarkable one was the restaurant of a Japanese-Sikh couple selling Indian snacks and foods. After their initiation into Sikhism, they had even adopted Sikh names - the man being called as Ranjit and his wife, Inderjit Kaur. The over-enthusism displayed by this newly-converted Sikh couple was simply over-whelming. Like a true Sardarji, Ranjit had long flowing lock and wore a turban tied around it. I was intrigued enough to enquire Ranjit as to how had he developed his fascination for Sikhism. He confided to me that he had worked in a restuarant at Chandni-chowk, Delhi, for quite a long time. He had learnt how to prepare several items of Indian food. Many Japanese who came to watch our performance crowded his Hotel and seemed to have genuinely relished his chicken samosa (¥ 250 each) and Roti chicken (¥ 300 each). A Video-recording of the famed Golden Temple of the Sikhs in Amritsar, was projected on a screen in his Hotel. In short, the scene provided a perfect ambience of an

Indian Bazar. Both sides of the banks of the Toki River were cemented on which beautiful murals were made. The natural beauty of the riverscape was complemented by the man-made murals and the two mingled to make it a single wide canvas. This instantly brought to my mind my own land. Even at the threshold of the 21st century, Manipur is unable to boast of a single mural. Why can't we mount at least one mural on the I.N.A. Memorial, Moirang? Why not one at the Gandhi Memorial Hall, Imphal? Another one at the Bhagyachandra Open Air Theatre is also long overdue. Sadly, we are lacking in leaders who possess artistic sensibilities.

This was our second visit to Tajimi City; we had visited it earlier in the month of August and gave a few performances. I was then introduced to an arts connoiseur named Madame Nakao. She had a private art gallery of her own. I made up my mind to visit her art gallery. Accompanied by her friend Bharadwaj Sharma, I went to see her price-less art collections. She possessed quite a number of them, mostly by French artistes. But there were also works by Picasso and Van Gogh. Madame Nakao spoke fluent French. Bharadwaj Sharma from India too spoke French. For about 5 years he worked in a Hotel in France. In Japan too, he was running a Hotel. He became a great friend of Madame Nakao, perhaps because both spoke French.

After visiting her private art gallery. Madame Nakao took us around for sight seeing. We visited the thousand-year old Eihoiji Buddhist Temple on top of Koke-jan Hill. The landscaping was stunning with waterfalls and fish-ponds. There stood an ancient tree (700 years old) inside the premises of the well-maintained and preserved

Temple. I was given to understand that it was the oldest tree in Japan and a show-piece for the tourists. We had to leave Tajimi the very next day for another City. So Madame Nakao came again at night to our camp to bid us good-bye. She presented each of our artistes with one Minoaki pottery for which Tajimi was famous in Japan. For me, she gave a present of a Minoaki Tea-set,the cups resembling small pots,which I still keep lovingly preserved at my home. It reminds me of Japan and its nice people.

Mrs. Kako Oyanagi, the chief organiser of our cultural programmes at Tajimi, gave us a farewell-dinner. It was a sight to cherish of Mrs. Oyanagi on that dinner night, clad in a pink Phanek Mapan-Naiba, (traditional wear of the Manipuri woman), pink Rani-phi, matching blouse and traditional Manipuri jewelry (Marei-Pareng). She was simply stunning and would be the envy of any Manipuri girl! Our female artistes had created this miracle. She had shown so much care and affection of our girls, that she had become almost inseparable from them, not leaving their company even for a moment. Since we were leaving the next day, she spent the night in the company of our girls, chatting and laughing. I could hear the noise of their laughter throughout the night and wondered what it was all about. Later I came to learn that she had been teaching a Japanese farewell song (Sayonara) that night to our girls. They rehearsed the song the whole night till the wee-hour of the morning till everyone became adept. The song runs like this :-

" It sumade-mo,

Tai-ru koto naku,

Tomodachi-de,Eyoh

Kyo no hi wa sayonara

Mata aawu he maa de,

Sin jiya-wu yoro kobi-o

Tai set uni siyo

aasuno he-o yume-meete

Kibo-no mitio."

(Freely translated, "I shall ever remain as a friend. Friend, good bye to you; but we shall certainly meet again).

When we presented this farewel-song at various places of Japan, they were pleasantly surprised and visibly pleased. They wondered how we had so quickly and correctly picked-up the song and could sing so well. I on my part, had nothing but admiration for the talent of the Manipuri artistes - how in one night they could master a song in an alien tongue.

On September 14, by 7 a.m., we were all seated inside the Bus and ready for departure when suddenly our artistes burst forth the Sayonara song and bade farewell to Tajimi. Our Girls wept inside the Bus, and our Japanese hosts too, they waved to us and gave a tearful farewell. We had to proceed to Yakumo village in Shimane Prefecture, a long 8 hours drive. On the way there was heavy rain. Driving along the Japanese Highways was a wonderful and novel experience for us - only my description would remain inadequate. Across the Highway, neon-lights flashed an umbrella dripping with drops of

Mrs. Kako Oyanagi in Manipuri Traditional attire.

water. This was an indication that the next city was having a heavy rain. We halted for sometime in a parking lot and had packed-noodles and a cup of tea as the day's lunch. By 5 p.m. we reached our destination viz, Yakumo village, famous for the Kumano Taisha Shrine of the Shinto religion. Besides, the place was believed to be the spot where fire was discovered for the first time in Japan. Yasuo Seko San, with his wife and daughter, had been waiting for our arrival; he was the chief organiser of the function at Yakumo.

Kumano Taisha Shinto Shrine

25

YAKUMO

After arrival at Yakumo, we hardly had any time for rest. A mobile telivision unit had arrived for a live-telecast of our dances by 6:30 p.m. Hurriedly, the artistes had put on their dance costumes for Maibi-Jagoi just in time to shoot for the TV programme. At 8 p.m. the people of Yakumo had done a great honour to us by arranging a reception for our group. After the reception, we had dinner together with our Japanese hosts and retired for the day.

During our long drive upto Yakumo, one memorable sight was the tea gardens of Japan. Our micro-bus passed through tea-gardens. Hundreds and thousands of iron-posts, about 10 feet in length, were affixed, cheek by jowl inside the tea-gardens. Those iron-posts were fitted with electric-fans for cooling the tea-plants. In the tea-gardens of India, evergreen shady trees are planted to provide shade to the young tea-saplings but I was quite amazed to see a hundred thousand electric fans over tea-plants in Japan. In a poor country like ours, there is not even sufficient supply of electricity for domestic-lighting, and more often than not, load-shedding is the order of the day. Think of it, the Japanese could indulge in providing one electric-fan for each tea-plant, while we are unable to afford a fan for a person!

The next day, Yasuo Seko San made a proposal to me. He suggested that we design a joint-programme with the Kagura dancers of Japan and the Maibis of Manipur. I informed my group of artistes about this proposal from the Japanese and all of them responded enthusiastically. We all went to the Kumano Taisha Shrine for a joint rehearsal. Seko san explained to us in detail about the proposed dance. He studied in India, in the Benaras Hindu University for about 2 years and hence could speak fluent English. Kumano Taisha Shrine had acquired a special significance for the Shinto religion, being the first place where fire was discovered - somewhat akin to the eternal flame at Andro in Manipur. Every year, during the Niwabishai Festival, large bonfires were lit around the shrine, situated just on the bank of a swift-flowing hill-stream. Colourful fishes are reared in the stream and visitors and tourists watched the fishes and fed them.

The dance number to be jointly performed by the Kagura dancers and the Maibis, was based on a mythological story of Japan. I experienced many instances of close affinity between the culture of Japan and Manipur. Just like ours the Japanese too , have their Maibis.They are called Miko in their language. Maibi-Jagoi is equivalent to their Mikono-odori. The gist of the mythological story on which our joint dance was based, is given below :-

Mikono-odori of Japan, similar to Maibi Jagoi of Manipur

The god Susano was driven out from Heaven by his sister for his wayward and wild habits. When he descended from Heaven, he came on the bank of a river. There he found an aged couple crying. Susano enquired the old man the reason for it. Thereupon the latter narrated their tale of woe. They had seven daughters. But the eight-headed serpent Yamatono Orochi had devoured six of them one after another. The seventh and the last surviving daughter, Enada Heme, also would soon become the food for Yamatono Orochi on that day. That was the reason why they were crying. Susano comforted the old man and his wife and assured that he would fight with Yamatono Orochi and slaughter him. The eight-headed serpent Yamatono Orochi was no ordinary serpent; it possessed super-natural powers[1] (Pakhangba?). Susano was no match for this mighty serpent. Within moments the serpent held Susano in the mortal grip of his powerful tail and attempted to devour Enada Heme next. At this critical juncture the latter began to dance gracefully. Captivated by the beauty and charm of her dance, the divine-snake stopped short of devouring the girl and Susano too, was released from his deadly grip. The story had a happy ending when Susano married Enada Heme.

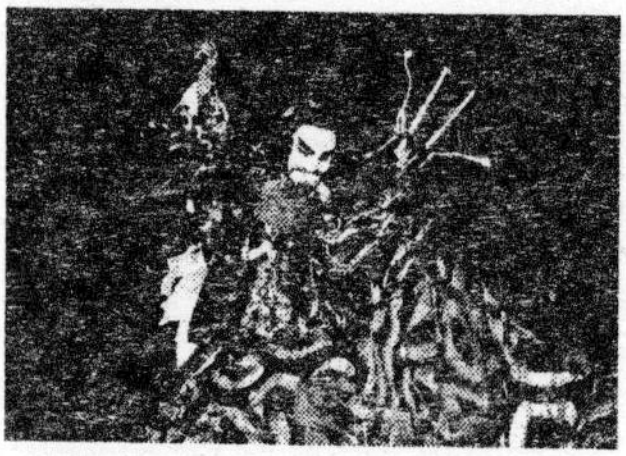

Fight bewteen Susano and Yamatono Orochi

---

1. Pakhangba – is the ruling deity of Manipur believed to assume the form of a serpent during day and that of a human being at night.

It was settled that the role of Enada Heme would be enacted by a Maibi of Manipur (Miss Usharani). It was such an exciting and challenging task for us. The Manipuri Group then arranged a joint rehearsal of the dance sequences with their Japanese counterparts. As per the story line, when Yamatono Orochi opened his mouth to devour Enada Heme, at that very moment the Maibi would burst-forth into dance Dhan Traka, Ta Khra Ka, Ta Khra Ka (3 minutes duration). The Kagura dancers of Japan played the role of Susano and Orochi. The rehearsal was a great success. The huge artificial serpent was also quite impressive-trailing its long tail measuring nearly 20 feet in length and 2 feet in diameter. The artiste playing the role of Yamamoto Orochi entered the improvised stage by dragging its long tail which was painted like a snake.

The next day was September 16, and our performance on that day in front of the Kumano Taisha Shrine was by far, the best and most satisfying experience, out of the 40 odd performances we gave in Japan. I still have no explanation why and how it turned out to be so. May be, we were blessed by the Almighty. The magnificent shrine, surrounded on all sides by lofty hills, the small hill-stream flowing silently in front of the shrine and our Lai Ikouba ritual starting from that stream ... ... all contributed to create an ethereal atmosphere. And when our Maibi performed the ritual of Lai Ikouba in the water of the stream and proceeded upto the Kumano Taisha

Meetei Maibi (Usharani) in the role of Enada Hime

Shrine by chanting invocations to the deity, the ambience created was a perfect Lai Haraoba in front of the Thangjing Temple of Moirang, Manipur. The natural surrounding and the breath-taking beauty of the hillscape contributed to the excellence of that day's performance making it a grand success. Words failed me to give expression to the pride and satisfaction which I felt on that day, when a Meitei Maibi as Enada Heme performed the Maibi Dance in the midst of the Kagura dancers of Japan. Unexpectedly, it turned out to be one of the best and most successful programme. People who watched it were all praise for the same. I shall be failing in my duties unless I record my admiration for Yasuo Seko san, the brain behind the innovative piece, for his creative talents. I had no doubt in my mind that the said programme was indeed, a fusion of the culture of the two countries. All the people who came to watch our performance sat on Tatami spread on the ground in front of the Shrine. The finale of our performance was Dhol-Dholok. The calm and serene backdrop of the Kumano Taisha Shrine reminded us of the Holi Pala of the Yaoshang Festival. All the Manipuri artistes appeared on the stage and started Thabal Chongba. Many from the audience also came over the stage and joined the Thabal Chongba and this unexpectedly turned out to be a most pleasurable and joyous ending of our performance for the day. The next day, i.e., the 17th September, we had to leave Yakumo village after staying there for three days for Fukuoka City to participate in the month-long 'Asia Pacific Festival'.

Through the kindness of Seko san, we were able to witness a Kabuki Theatre that night at the nearby city, Matsue. In Kabuki Theatre, there are no female artistes; their roles are played by male-artistes which instantly

Kabuki Theatre

brought to mind our female impersonators of Shumung Lila. Believe me, Manipuri male artistes of Shumang Lila in female roles are in no way inferior to their Japanese counterparts of Kabuki Theatre.

Since we stayed at least two or three days in Yakumo village, we had an opportunity to observe the Japanese style of paddy cultivation. The paddy crops were then ripening in the fields of Japan. I wish to share my interesting experience of watching a Japanese couple performing the entire process of harvesting, thrashing, packing, and transporting the product of about one Pari (one Hectare) of paddy field within a few hours. From my own experience, I know that at least 30 to 40 labourers would be required to complete the process of harvesting a Pari of paddy field in one day. Even with the help of so many farm-hands, it is still doubtful if the job would be accomplished at all in a single day. It was incredible (at least to us) how a Japanese couple, with the help of a few machines, had completed the entire process of harvesting a large field in one day. We watched the different stages of the operation with amazement and admiration. They used a very small harvesting machine about the size of a tri-cycle for children. The man started the machine and pushed it in between rows of neatly planted paddy. It might not be an overstatement to call the machine a mini-marvel. The small machine simultaneously performed the functions of harvesting the paddy, binding them into small sheaves by thread and hurling them on one side. A bundle of thread was hung on the machine. In this way, it took hardly half an hour to harvest one acre of paddy. At one corner of the field stood a paddy-thrasher, its engine running all the while. The machine was not very big but its functions were numerous. The sheaves of paddy were collected by his wife and fed into the thrasher which instantly threshed them while the grains emerged from one side. Those grains of paddy were collected in small hessian bags and as soon it was full, his wife zipped its mouth. Instantly, another hessian-bag would automatically drop and the same process was repeated till the entire operation was complete.

While paddy threshing continued on one side of the machine, it simultaneously cut the hay into small pieces, about one or two inches in length. The Japanese farmers did not gather the hay after harvesting the grains from it. Those small pieces of straw were scattered all over the field to serve as manure. While in Manipur, we perform every stage of cultivation manually, in Japan a small machine could substitute almost all the manual labour. By evening, all the paddy bags were loaded in a small pick-up van and the couple drove away. In this way, harvesting one Pari of field, threshing the paddy and transporting the produce upto one's home was completed in one day.

On the morning of 17 September, we bade farewell to Seko san and his wife and other friends of Yakumo and seated ourselves inside the Bus ready for departure. At that instant, our artistes burst-forth the farewell song of the Japanese, 'Sayonara'. On hearing that song, Seko-san and his wife, with tears running down their cheeks, waved their hands and bade us good-bye. This prompted our artistes to cry, some of them openly inside the Bus. Our Bus journey on that day would take about seven to eight hours and Seko-san had most thoughtfully provisioned the Bus with fruits, snacks and other eatables. We left the lovely village Yakumo and its immortal shrine, Kumano Taisha and headed for Fukuoka City.

Yasuo Seko san leading a group of Manipuri and Japanese in Thabal Chongba Dance.

26

FUKUOKA - II

On our way to Fukuoka, there was a strong wind and the temperature, suddenly dropped. By that time we have received reports of widespread damage in Tokyo City caused by a typhoon. Although the fierce wind would not reach our place due to its long distance from Tokyo, the weather had undergone a sudden change for the worst and we felt very cold. We reached Fukuoka by 5 p.m. and stayed at Genyo Civic Hall. This was our second visit to the city; the first being on 28 August when we came here to participate in the International Ethnic Festival. During our second visit, we stayed for 7 days in this city and performed Manipuri Dance daily.

Genyo Civic Hall, where we were put up, was situated just near the sea. Hardly 200 feet from the Building and we could touch the sea. The soft murmur of the waves lapping the shores, its intermittant sound of breaking on the sea-shore at regular intervals provided a rhythm while the hooting of siren from a ship anchored at a distance interrupted the stillness of the night. It was entirely a new experience in my life. Everyday early in the morning, I strolled bare-foot on the sands of the sea beach for some time and then stepped into the water of the sea for a while. Our boys and girls joined me in the walk on the sea-beach. All of us felt an entirely new pleasure, like a child with a new toy. Coming from a hilly place like Manipur, where the sea is non-existant, our response was only too natural. Before this, I have already had acquaintance with the sea - at Orissa, Goa, Bombay etc. but never had such a wonderful opportunity of staying for seven days at a sea-front and watching the mysteries of the sea and that too, in Japan.

The Asian Pacific Festival was organised every year for one whole month in Fukuoka City, designating this month as 'Asia Month'. In this Festival, artistes from various Asian Countries were invited to present their traditional cultures. Since no other group from India came here, we had the privilege of representing our country in that prestigious Festival. The performance stage was constructed with ordinary materials like bamboo and wood but the design, both artistically and aesthetically, was superb. Unpolished whole logs of Pine were used while making the stage which they had christened as 'Asian Soul'. The stage was constructed on a lawn facing the fifteen-storied Fukuoka City Hall. No seating facilities for the audience were provided. Instead, a number of eating-Kiosks were opened by Asian Countries in front of the stage, but surprisingly, none from India. The nearest to Indian-like delicacies, were served by the Kiosk of Nepal.

At night there was laser-beam display. All visitors to this Festival could avail of free international telephone call facility. Naturally, we telephoned Imphal almost everyday! The performance of the Australian Aborigines highlighting their traditional cultures on 18 September, just ahead of our own programme, was most exciting for us. I wish to share my experiences and the thrills with the readers about the cultural programme by the Aborigines. In their first item, they demonstrated their traditional method of making fire. A piece of dry twig was held in between the palms and fixed on a piece of wood. Dry tinder was placed at the point of contact between the stick and the wood. With the help of the two palms, the wooden piece of stick was swiftly rotated over the dry wood for about three or four minutes. The friction caused by the two pieces of dry wood heated the tinder and smoke emitted from it. The Aborigine then collected the smoking tinder in his hands and made repeated puffs by his mouth and behold! Flames lept up from the dry tinders within seconds.

The musical instrument which is uniquely for the aborigines is the DIDGERIDOO, locally known as YIGI YIGI. It is considered to be the oldest musical wind-instrument in the world. The aborigines demonstrated their skill in playing that musical instrument, which was a simple termite-eaten hollow piece of Austrialian eucalyptus, about four feet in length and three inches in diameter. The aborigines go to the forest in search of the right hollow branch. Through generations of acquired skill and eyes, only could they identify such a hollow branch. *They blow this hollow instrument to

---

*These termite–eaten Australian encalyptes, are being crafted skilfully and painstakingly by skilled Aboriginal craftsmen to make the instrument.

produce a variety of sounds. Depending on their skill and regular practice, the aborigines are capable of producing the sound of different brids and animals. On the day of their performance, they imitated the sound of a Kangaroo and an Emu's call. We were all familiar with Boomerangs; everybody had seen it in picture-books and TV It is a curved and flat woden missile used by the Australian aborigines for hunting purposes. But we had never seen how it was handled by the aborigines. They demonstrated the same in a near-by Park. The use of the boomerang is an art in itself which can only be perfected by continuous rigorous practice. Any and everyperson cannot just pick-up the weapon and fling it properly. The boomerang appears to be a simple and uncomplicated weapon at the very first sight, yet it is highly effective and an amazing one. In the hands of a skilled thrower, the weapon after making a flight for some distance, would return to the thrower. When the boomerang turns its flight back towards the thrower, it is caught by the palms of both the hands. In the hands of an inexperienced thrower, unless he is careful, the weapon would often strike the thrower himself causing injury. This precisely is the reason why the word boomerang is figuratively used in English literature to mean' an action or remark that cause unexpected harm to the person responsible for it'. It is also known as Wangal among the aborigines. According to their customs and traditions, it is a taboo for women to touch the weapon. From interactions with the leader of the Aborigine Group, (his name was Willie Brim) it was learnt that he belonged to TJAPUKA tribe of the aborigines of North Queensland, Australia. He further informed me that the aborigines spoke more than 300 different dialects.

For the next three days, as a precautionary measure against an impending typhoon, all the programmes earlier scheduled for the 'Asia Month' were cancelled. Since we had some free-time at our disposal, I paid a visit to the Fukuoka Museum. I was particulary keen to have a look of the Golden Seal which was declared as a National Treasure of Japan. Unluckily, we could not see the exhibit as it was displayed for viewing only on certain specified days. However, I had the opportunity to view other galleries of the Museum.

The next memorable art-pieces I saw that day were the four bronze statues erected in front of the Museum. Those four statues were sculpted by the eminent French artiste, Emille Antoine Bourdelle (1856-1929), who was a student of Rodin. Later on, she had evolved her own style of sculpture and perfected the same in her later works. The four statues were named as (1) La Liberte, (2) La Victoire, (3) La Force and (4) La Eloquence. Those statues were completed between 1918 to 1922. They measured 375 cm. high and 110/120/130 cm. in diameter. The statues were first installed in the Argentine capital, Buenos Aires. On the occasion of the Centenary Celebration of the founding of Fukuoka City, the statues were brought from Argentina and installed in front of the Fukuoka Museum. The sight of those grand statues brought to my mind thoughts of distant Manipur where I live. The Centenary Celebration of the Battle of Khongjom (the Anglo-Manipur War-1891) ended with only a few lack-lustre programmes. No lasting monument, either in the form of a statue or even a centenary hall was raised. It was a missed opportunity since centenary Celebrations are extremely rare occasion which comes only once in the life of an individual. No wonder, no sooner had the celebration was over, the historic last war for freedom of Manipur also faded from people's memory.

During our stay in the Genyo Civic Hall for about a week, we shared our meals with the Japanese people. Many of them who came to visit us during dinner and lunch were invited to join us to have food. They genuinely seemed to enjoy the hot spicy Indian curry. They were very keen to taste the Indian cuisines and came in hordes during dinner, but never did they come empty-handed. Each of the visitors brought lots of eatables as presents for us. Some had brought potatoes from their own gardens, some brought rice, fish and fruits etc. The Japanese normally take their food with chopsticks (which they call Wari basi), but on that day they too ventured to eat with their fingers as we Indians did. During dinner times, streams of Japanese visitors - sometimes as many as ten guests - would come and partake the dinner which we cooked. We on our part would ensure to cook curry daily in excess of our requirements in anticipation of any unexpected arrival of some Japanese guests. It was such a pleasure and fun for us to watch our Japanese friends eagerly tasting our food and eating with their fingers. Come to think of it, they were not poor and destitute people asking for food; they came simply out of their curiosity for our cuisines. I am reminded of an unforgettable incident in Fukuoka. Many of the

Shri Kuldip Sahdev, Indian Ambassador in Japan and his wife, with the Manipuri Artistes

performing places were located far away from our lodge. And more often than not, we used to carry pre-cooked dinner with us in our Bus, lest we might not find time for cooking after our performance. Normally, we had our dinner immediately after our performances in an adjoining room. One night when we were about to start our dinner, a young Japanese girl (about 24/25 years old) and a middle-aged man entered the room wherein we were taking dinner. They came forward and asked whether I would permit them to taste a little of our food. I had never heard in my life such a sweet and unsophisticated request. A sudden surge of emotions of love and respect for this wonderful race overpowered my senses. I was completely bowled over by their sincerity and frankness. Which race in the world could say like this except the Japanese? This brought to my mind the behaviour of the Manipuris as a sort of contrast. We have a strange habit of refusing food when first offered by someone, yet taking the same without even slightly offering a word of thanks. The simple request of the Japanese to taste our food won over our hearts. We offered our rice and curry in a plate and how thankfully and happily they partook it. Their food were of a different kind; hence they were keen to try our hot and spicy food. They were not to be mistaken for poor people begging for food; all of them came from well-to-do families. That was precisely the reason why they had the confidence to ask for food without any inhibitions.

It seemed that the Japanese people hardly had the patience even to take down the cooking pot and the food from the oven. They start eating straight from the oven. It was a common sight; they picked up the steaming hot foodstuffs from the oven with the help of chopsticks, brought down the temperature by repeated puffing with their own breath and gulped it instantly. It was quite a strange spectacle for us. Why can't they remove the fry pan and cool it down for sometime before taking it? Perhaps because of this bad habit of eating steaming hot food, most Japanese have bad teeth. It seemed that they did not take enough dental care. While brushing their teeth in the morning, I did not see them using tongue-cleaners as we did. In Japan, bicycle is used widely, both by males and females. Being a cold country, their cycle handles were fitted with wollen gloves to ward-off the cold. That was quite a novel experience for us and we, in Manipur, also could take a lesson from them in this regard. While staying in Fukuoka, we had many memorable and amazing incidents. Just near our lodge, there was an underground Railway system (a Metro Station). One of our artistes, a Pung Player requested me to take him to the Metro. I told him that such under ground Railways could also be seen in Calcutta or Delhi and promised to take him there on our teturn to India. He persisted and pestered me. I yielded and early next morning only two of us came out from our camp for a morning stroll and see the Metro Railway. There was no one around; the station seemed deserted. We descended the stairs and stood on the platform there. Some school going children however, started trickling. No railway guards or ticket-sellers also were around. The school children pushed-down a card or ticket into a box kept near the embarkation area of the platform. This emboldened us and we ventured towards the unmanned platform. But no sooner had we ascended the platform, an iron-bar suddenly thrust out, Heaven knows from where, and blocked our passage. The meaning was mute but clear- we were debarred from entering the plateform without holding a ticket. There was nobody to operate that contraption, everything was automatic and electronically controlled. Both of us were caught completely by surprise and was transfixed on the spot. How funny and embarrassed we felt! We turned around to see if anyone had been watching us. Ignorance; Thy name is foolishness.

During our stay at the Genyo Civic Hall of Fukuoka, we developed an intimate friendship with Shiro Ouchi San, the officer-in-charge of the Hall. He took us driving in his own car for sight-seeing around Fukuoka City. His small car (like our own Maruti) was ten years old and told us that he would soon be disposing-off the same for 5000 yen (equivalent of approximately Rs. 2,000/-) He further stated that disposing-off such second-hand cars in Japan was a big problem; people who purchased used-cars were rare.

A car manufacturing company had agreed to buy it for a paltry sum of 5000 yen on condition that he should buy one of their new cars. In Japan, space for car garages were another problem since their houses and homestead lands were extremely limited. After his wife and daughter had parked their cars, there was hardly any parking space left for our friend Ouchi san. That was the reason why he kept his car in the Genyo Civic Hall where he worked. Another problem in Japan was that there was no space for dumping old and worn out cars. One had to pay heavy taxes for dumping too. Old cars were crushed by heavy machines to retrieve only the iron. For this also, they had to pay a fee of 5000 yen to the Government. He therefore, told us that he was prepared to present his old car to us if we cared to take it back to India. His ten year old car was fitted with air-conditioning facility and as good as our new one. Some of our artistes were quite tempted to have a try; only I dissuaded them.

Shiro-Ouchi San told us that in the coming weekend, he would be going to Korea to play Golf. One of his friends there had invited him and sent an air-ticket also for him. This was quite a surprise for us to hear. Why should he go to Korea just for the fun of playing Golf ? Why couldn't he play in Japan, if he were such an enthusiast of the game? Ouchi San enlightened me on these baffling questions for me. It was very expensive to play in Japan. For a day's play, he had to shell-out at least five lacs yen. On the contrary, it was much more economical to fly to either Korea, Indonesia or Malaysia in the week-end and return to Japan on Monday, just in time to attend office. The most popular sports of Japan are Baseball and Golf. We hardly come-across people playing Football, Hockey or Cricket. But wherever we went in the towns or cities of Japan, we had seen grand Baseball Tournaments being held there. Being an affluent country, they could afford all the luxuries of life which we fail to even imagine. It was also an indication of their high spending capacity. Ouchi San familiarised us with various faces of the city. He showed us how Turkish Baths were operated in Japan. There was government restrictions on call-girls practising their trades openly. Hence they invented an alternative institution called the Soap Land to by-pass the law. He took us to some Soap Lands of the city. In a narrow lane, there were rows of small shops. There the

Kabutogani, a rare crab found in Japan

call girls treat the clients with massage and hot-water baths. And for an hour of such a treat, the Soap Land girls would charge six thousand yen per customer (100 yen was equivalent to our Rs. 30/-). Basically, Soap Land was hardly any different from our Red-light area; they merely gave an elusive name to it. Shiro Ouchi San presented me two very rare species of Japanese sea-crab (dried). This particular species of crab was not found anywhere in Japan except in the sea around Fukuoka and Okayama. It is called Kabutogani in Japanese (Scientific name; Tachypleus Tridentatus). It is rounded in shape; its entire body was thorny and possessed long tails. He acquainted us with interesting ancedotes about this crab called Kabutogani. This particular crab is always found moving in pairs - a male and a female. They are never separated form each other. The female Kabutogani is larger in size and possessed a longer tail and larger limbs from their male counterparts. The male Kabutoganis always remain close to the females, never leaving her alone for a second. They are idlers and rode on the back of the females and swim together. Hence, in Japan, a hand-pecked husband who always remained clinging to his wife is nicknamed as Kabutogani! The dried Kabutogani which I had brought from Japan is now displayed in the Natural History Gallery of the Manipur State Museum.

Our days in Fukuoka were coming to an end. From here we had to proceed to Yokohama. On September 25, we bid farewell to our good friend Shiro Ouchi and left Fukuoka. The drive from Fukuoka to Yokohama was a long one and it was not possible to cover the for bidding distance in one day. We therefore halted one night at Ako village where we had stayed earlier. Our micro-bus running at a speed of 120/130 kms per hour reached Ako at 7 p.m. In Japan, vehicle toll taxes were very high. On entry to every city, a vehicle had to pay taxes. We halted at many places for payment of toll taxes. On September 26, we reached Yokohama at 6 p.m.

The writer at Yokohama port with Drummers - Rajmani and Boinao.

27

YOKOHAMA

It was our second visit to Yokohama; our first visit to this city was in July when we took part in the Summer Night Festival. During our second visit, we stayed for one week in the city which afforded us ample opportunities to visit Tokyo and some other places near-by. There was a Public Park not far from the place where we stayed. It was a place frequented by a good number of morning strollers and we also did the same. In Japan, majority of the people who came in the morning for exercise and jogging were elderly persons. Hardly any young men or women could be seen performing these early morning rituals. Most of the early-birds came with their pet dogs which were a craze among the Japanese people. Those who visited the Park with their pet dogs invariably brought a plastic bag and some old newspapers with them. The pet dogs were coaxed by their owners to relieve themselves on a piece of old newspaper. The owner then collected thc animal excreta in the plastic carrying bag for disposal in an appropriate waste-bin. The Park was not to be littered with animal-dung; it was the responsibility of the owner to clean the same. It brought to my mind the old Manipuri proverb "when a dog becomes wild, hold its owner". The same was true, although in a slightly different context, in Japan. This was also a glowing instance of the high civic sense of the people. Another interesting and somewhat surprising sight was that the pet dogs were pampered by their owners to a ridiculous extent. They treated their pets as if their own children. Whether it was due to over-pampering or our first exposure to such experiences in life, I couldn't say; but it was truly a bizarre sight for us. The owner purchased an ice-cream; she lapped it first and then offered it to her pet-dog to lap it. In this way, the whole ice-cream was lapped alternately by the dog and its owner till the last drop. Some of the Japanese women dolled-up their pet-dogs like their babies - ribbon on the head and butterflies decorating the same.

A very distinctive Japanese trait which I had observed was that they talked in very soft and low tones, almost in whispers. Loud talks in public places was considered impolite. To an Indian like me, whose ears had been attuned to the shrill voices of the Biharis and the Bengalis, I could not but admire this trait of the Japanese. Even their young girls would giggle and laugh in very subdued tones. I had also observed with dismay, another sight in the street corners of Japan. There were a number of pornographic literatures and pictures being sold on the wayside. Young students slinging school bags on their shoulders would browse through the porn materials for hours and nobody raised an eye-brow. Japan is considered an orthodox society, very proud of their traditional culture and high moral values. But they tolerated their young boys and girls indulging in pornographic literature and pictures. This is not possible in India where the police would arrest the sellers. On 29 September, we were informed that His Excellency the Ambassador of India in Japan, Sri Kuldip Sahdev and his wife would be pleased to host a tea-party for the visiting cultural Troupe from Manipur. On this very day, the speaker of the Manipur Legislative Assembly (W. Nipamacha Singh) also happend to be in Tokyo. The reception, to be followed by dinner in the Indian Embassy, was scheduled on the next day at 7.30 p.m. Although the reception was at night, we left Yokohama after an early lunch for Tokyo at 11 a.m. in order to avail of the opportunity of sight-seeing around Tokyo. The drive from Yokohama to Tokyo took only about one hour.

At 2 p.m. we reached Ginza Street, arguably the most well-known place of Tokyo. It is an exclusive area for the wealthy and affluent section of the society where the land-price is sky-high. How many million yens one had to shell-out for a square feet of land at Ginza is simply mind-boggling. There were many Government Offices located in the Ginza street including the Tourism Office of the Government of India. We called on the Director of Tourism, Ms Purnima Shastri and had some discussions. Personally I felt that there was hardly anything special about Ginza street. It is somewhat like our own Connaught Place, New Delhi. The only difference was that the cost of land was much higher in Ginza than in Connaught Place. And there was nothing surprising in it. Being a small island full of hilly areas, Japan is awfully lacking in flat level lands. After strolling for a while in and around Ginza, we headed for the Renkoji Temple where the ashes of Netaji Subhas Chandra Bose were being kept. The head of the Temple-designated as Chief Minister of the Temple, Rev. KOSHI MOCHIGUKI arranged for us to view the copper-casket containing the ashes of Netaji and pay our respects. We offered flower and incense sticks there. The Priest presented me a book on Netaji written in Japanese language.

Taking advantage of an Exhibition on Indian Folklore held in Tokyo, we went to visit the said Exhibition. When it was done with, we went to call on S. Mismi San, Director-General of Indo-Japan Friendship Association. He gave a tea-party for the Indians from Manipur (in Japan, we were always addressed as Indians, not as Manipuris). Incidentally, I am also a member of the Indo-Japan Friendship Association. I handed over some books on Manipur to S. Mismi San presented by the Secretary. Indo-Japan Friendship Association, Manipur Branch. He was very pleased and thanked me profusely for the same. I had a discussion with him on the role of this Association to strengthen the Indo-Japan Friendship and what other steps could be taken towards that end. I told him that we had come to Japan and shown to the people a glimpse of the rich cultural traditions of Manipur. In like manner, we would very much appreciate if he could make reciprocal arrangements for showing the art and culture of Japan in Manipur. That would go a long way in furtherance of the aims and objectives of the Indo-Japan Association. He endorsed my feelings and sentiments and suggested me to send a proposal on reaching Imphal.

While meeting these various personalities, it was already 7 p.m. and the time for going to the Indian Embassy had approached. We went straight to the Embassy where we were warmly received by the Indian Ambassador and his gracious wife. Except for a handful of his close friends and associates, there were not many guests at the party arranged for us by the Indian Ambassador. Among the Japanese guests present, there were quite a few of them who had already been to

Manipur like Mr. Sei. The few assembled guests were enthralled by the cultural programmes of Manipuri dances and songs. The most memorable moment was the finale of the night's performance when His Excellency the Ambassador of India, Sri Kuldip Sahadev and his wife joined Thabal Chongbi with the Manipuri artistes.

The function was made more enjoyable and fulfilling when almost all the guests also participated in the Thabal Chongba. Performances over, it was now time for dinner. The menu served on that night was equally unforgettable. After a long and continuous journey throughout the length and breadth of Japan and having been deprived of any Indian food for a long spell, the oppopturnity to have lavish and mouth-watering Indian delicacies like Paneer-Pulao, Chicken Curry, Raita, Dal fried in butter, Nanki Roti, green vegetables subji etc. etc. was simply a heavenly experience. After the dinner, the guests were treated with sweets like gulab Jamun and Rassogollas. O, what an incredible amount of food the Manipuris devoured on the dinner night hosted by the Indian Amnassasdor and with what relish! As for me it was an embarrassing moment! What would the Ambassasor and his wife would think of us!

After bidding farewell to the Indian Ambassador and his spouse, we returned to Yokohama at 11:30 p.m. Hearing that some artistes from Manipur had arrived at Yokohama, a young Manipuri youth (Rajesh of Yaiskul) working at Japan had been waiting for us. This youngman was an Electronic Engineer working in a Computer firm there. I felt proud and happy that a Meitei was working in Japan from distant Manipur. Incidentally, he was the only Meitei working in Japan. Unlike the U.S.A. or UK. it was not very easy to get employment in Japan by foreigners. They did not encourage foreign workers in their land. That was the policy of the government. Instead, they build-up their own manpower in diffrent fields of science and technology. Young Japanese were deputed to various countries for research and training in higher studies with the sole purpose of serving their motherland on return. The Manipuri saying that 'better groom your own brood rather than backing on someone's good son' is not a proverb but a reality in Japan.

On October 1, an exceptionally creditable performance of Manipuri Cultural items was presented at Tokyo. Our Ambassador too, graced the occasion. It could safely be said that the performance at Tokyo was the most well-organised and elaborately arranged show. The guests represented almost all the elite and gliterrati of Tokyo City. Our performance was organised by a voluntary organisation called "Namaste India" at the Sumaida Riverside Hall. The number of Indian Restaurants and stalls selling Indian ware provided an ambience of an Indian Bazar. There was heavy rush of people buying these Indian goods in great gusto - some buying Salwar Kamiz, Sari and Bangles etc. crowding almost every stalls.

'Namaste India'

On one side, there were rows of Restaurants and eateries selling ethnic Indian foods like Tandoori Chicken, Tandoori Roti, Naanki Roti, Samosa, Pakori etc. I was curious to see how they prepared Tandoori Roti and Chicken there. Their Tandoors were made of steel, not of mud as is the practices in India. Another interesting feature which I noticed was the Indian names given to the Restaurants such as Lumbini, Gandhara, Moti Mahal etc etc. The food-items sold there were greatly relished by the Japanese people. Some of the Hotels had such a rush that the customers had to form a long "Q" in front of them.

Before the start of our performance, a Japanese lady entered our dressing-room to meet our girls. She spoke good English and Hindi, too. Mrs. Chihiro Khoiso had been living in India, learning Hindi in Wardha (M.P.). There she had made friends with some Manipuri girls. She said she loved the Manipuris and had even learnt to sing a Manipuri song from her Manipuri girl-friends. "Then let us hear a Manipuri song which you know", I told her. With a shy smile on her lips, unexpectedly, she burst-forth an all-time hit of Pahari viz.

'Khomlang Laman Shinggamloy Nanggi
Nungshi Chaobi Hey Ema,
Chingda Kaduna Shingjanba
Pat-ta Kumduna Long-khonba ...'

(This song is an ode to Mother and Motherland. Freely translated : ' it is impossible to repay for your breast-milk. O, loving mother ! How you go up the high mountain for cutting fuel-wood; how you go down into deep lakes for fishing ...')

A Japanese girl singing a Manipuri song ... and that too, at distant Tokyo. O! What a pleasant surprise. Ms. Khoiso had indeed charmed the Manipuri artists who were no less thrilled. She had obtained the degree of Doctor of Philosophy from Pune University. When I reached Manipur, she sent me a letter requesting to find out the whereabouts of her friend Thambaltombi of

Khurai and send her complete address. It was quite a Herculean task for me to hunt for a certain Thambaltombi; there are hundreds of girls bearing similar names in Khurai. It was like searching for the proverbial needle in a haystack. After a long and arduous search, I was successful in locating Thambalatombi, the friend of Chihiro Khoiso of Japan. What a sense of satisfaction for me. Two long-lost friends, one in Manipur and the other in Japan, were brought face to face and revived their friendship by fresh communications.

That day at Tokyo, we had the luxury of having our favourite Indian dishes which were available in plenty there. We were served with profuse quantities of Rice, Subji, Mutton Kabab, Tandoori Roti....etc. and everybody's spirit was quite enlivened. After food, we had to return the same evening to our station at Yokohama. The main reason was that there was no place to accommodate us; hotel charges were exorbitant and we simply could not afford it. If nineteen of us had to stay one night at a Tokyo Hotel, we had to spend nearly Rupees one lac. That precisely was the reason why we had to stay wherever the organizers could manage within their limited budget.

On October 3 and 5, our performances were scheduled at two tiny islands in the Pacific Ocean. We were prepared for that. There are six small islands in Tokyo Prefecture viz, Oshima, Toshima, Miyakojima, Hachijyojima and Aogashima. Our first performance would be on October 3 at Hachijyojima Island.

Hachijyo Jima Island

28

HACHIJYO JIMA

To reach Hachijyojima, one has to go by ship from Tokyo Harbour. We left Yokohama by 8 p.m. and reached Tokyo Harbour by 9 p.m. We were on board the ship bound for Hachijyojima. Hasegawa-san and his young son, Munna accompanied us. A representative of the Japanese Broadcasting Corporation, the NHK (Nippon Hoso Kyokai) also joined us. Keito Hashimoto-san was a youngman of 24 or 25 years; an ever-smiling pleasant companion. He shared our meals cooked by our girls. The trip to Hachijyojima gave us the opportunity of a sea-voyage for the first time in our lives. The huge ship cruising in the sea, breaking the giant waves and the tranquility of the night awed and thrilled us as had never before. Sitting on the deck of the ship and watching the turbulent waves in the sea was totally a novel experience. I feel a slight dizziness, probably because of the shaky ship. Throughout the night, the ship continued sailing and the next morning at 9.30 a.m., we landed at Hachijyojima Harbour on October 3. A number of people were there at the harbour to receive us and they volunteered to take us in their cars upto our lodging. Hachijyojima is a small island; it is about sixty kilometers in circumference. It has a small population of only about ten thousand people. The island was formed by solid lava from volcanoes thousands of years ago. Even today, there is one active volcano, spewing fire and smoke from its crater. The entire land mass of the island was filled up with black rocks; hence unfit for any vegetables or paddy cultivation. The main occupation of the islanders is fishing. Another source of revenue is Tourism. The surrounding sea-scape was enchanting - the blue seas encircled the island on all sides. Many surf-riders and tourists were frolicking on the surface - some had brought their water-skis, some were rowing in their small boats. At such an isolated island, we ran upon two American girls. They were English teachers. Nowadays the younger generation in Japan had started learning English avidly. Earlier, learning English was never a priority in the Japanese education system and as such there were not many Japanese who were fluent in English. It appears that the Japanese have now realised that without the knowledge of English they would not be able to catch-up with the modern times. In this island, there was not a suitable Hall big enough for our performance. Hence, it was arranged in a Gym temporarily for the day by putting steel chairs on plank floor. (A word about the numerous gymnasiums abounding every nook and crany of Japan. Each and every small town or a village possessed a gym. This is indicative of the active government support towards the development of games and sports in the country). A large numbers of people came to witness our performances and we too did our level best to present a successful programme. The Lai Haraoba Dances, and that too, in the middle of the Pacifice Ocean ... "Chekla Paikharabana Pombi Hanjillakpada/Cheklagi Kaidongpham Khangdrabana, Pombi Kakngao Naredo/

Shabi Inemacha Pammubi ..." (Freely translated: the bird has flown: when it returns there is nowhere to perch. The bird is confused and confounded O, my beloved). The high-pitched notes of Manipuri folk songs accompanied with the Pena music reverberated the wide expanse of the ocean. That was my greatest fulfillment and pride. Alone I pondered; we had brought the culture and traditions of distant Manipur to the middle of an ocean in Japan which would be hard for any group from Manipur to repeat. We had a rare chance to highlight our culture and the people of Hachijyojima savoured it every moment. What more satisfaction one can expect from such an experience? We had been amply rewarded for our visit to Japan; the culture of Manipur touched every nook and cranny of the country.

After the performance was over, it was unthinkable that the thoroughly exhausted artistes would have some respite, had dinner and take rest for the night. Wherever we performed in Japan, it was customary, nay, almost a ritual, to attend a social gathering followed by a party after every show. On days of hectic programmes, their social gatherings and parties were almost a curse for us. The very thought of participating in the social gathering after each performance made us sick and sometimes angry, too. But since we had come here crossing seven seas, we had to oblige our guests with utmost reluctance for the sake of decency and decorum.

A live Volcano at Hachijyo Jima

And we suffered silently throughout the party, never enjoying for a moment. In Japanese parties, their most popular drink is the "Ocha", which is nothing but plain tea without sugar or milk. It was followed by beer which they gulped down like fishes. My host at Hachijyojima presented to me a CD cassette recording of the sounds of many kinds of Japanese birds in different seasons. It was a rare cassette solely recording bird-songs. I played this CD casette to a select few audience in Manipur who appreciated it very much.

We had been making arrangements for giving a performance the next day in another small island called the Aogashima, meaning 'The Blue Island'. The programme at Aogaahima Island was important and extremely significant for me. None from Manipur had ever given performance here nor was there any possibility for any performing group to do so in future. Precisely for this reason, Sri Ratan Thiyam, who had organised this Cultural Troupe, also flew from Imphal and reached Hachijyojima. Tomorrow, we would be going with him to Aogashima for performance there.

29

AOGASHIMA

The area of Aogashima Island is a bare 9 square kms. From Hachijyojima we flew by Helicopter which took only about 15 minutes. It was a small chopper and as such, we were ferried in two batches. It is a tiny island habitated by fourteen, I repeat, fourteen families only, consisting of a mere 200 souls. From aboard the Helicopter we could see a cluster of some five or six houses which instantly brought to my mind the vision of the tribal villages of Manipur seen from the aircraft. Is there any Manipuri, returning home after a long time, who could resist the temptation of looking down and watching from the aircraft the hillscape below? O, look here, there is Imphal! Small tribal hamlets on the hills gradually rolled in our view. Suddenly, a feeling of intense love and yearning for Manipur surged in our hearts and we felt home-sick. Aogashima Island was formed by solid volcanic lavas in the course of thousands of years ago. Viewed from the air, it

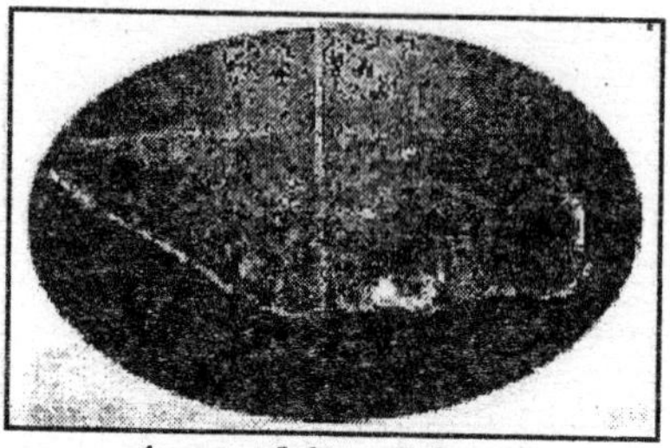

Aoga-shima Island

resembled a small hillock, abruptly popping-up from the sea. The village is perched high on the plateau made of solid lava.

On our arrival at the Island, the organisers led by Hasegawa san received us. None of them could speak English except one lady teacher. One among them, Hiuchi Akinari, could speak Hindi tolerably well. In 1992-93 he worked as a volunteer in Mother Teresa's Missionary of Charity in Calcutta. It was a pleasant surprise to meet a Hindi-speaking Japanese and that too, in a God-forshaken island in the Pacific Ocean. The means of livelihood and the life-style of the handful of people living at Aogashima, was, intriguing. In this wind-swept hillock of an island, nothing was availavble; the island produced nothing. I could see a few cattle and people doing dairy farming. But that was not enough to sustain the economy of the islanders. The entire population was virtually fed by Tokyo; everything required by the islanders were provided by Tokyo at heavily subsidized rates. In certain years, when there was scarcity of drinking water supply, the same was transported by ship from the mainland. There were two school teachers in this island. Upto 50% of their salaries, even their House rents, were borne by the Government at Tokyo. My reader might be surprised, and I too was no less so, as to why Tokyo had to maintain at such a fantastic cost these 200 odd people of Aogashima.Why were we invited to perform in such a non-descript island for the benefit of only a handful of people? The answer to these various questions would soon be unfolded shortly.

Our natural reaction and query was why can't those 200 people be shifted and re-allocated elsewhere in a suitable place ? No, that was not to be. The inhabitants vehemently resisted any move to leave the island, where their forefathers had lived for generations, and lost their lives in tragic circumstances. There was a poignant and emotional story behind their stiff resistance and that itself was the history of this 'Blue Island' - Aogashima. Thousands of years ago, there was a volcanic eruption in this island. All the inhabitants made frantic attempt to escape from the island to a safer place. It was impossible to accomodate the entire population in one available ship and many were stranded in the island doomed to perish. One would well imagine what should have been their fate. Everyone perished in the hot molten lava from the volcanic eruption. The present inhabitants of the island were the descendants of the original settlers who lost their lives in tragic circumstances. To them, the island is a visible memorial to their ill-fated ancestors and hence they vowed never to leave the island. They refused the persistent proposals of Tokyo administration to shift to some other more hospitable area. Since there was no scope for any gainful economic activity in the island and no other viable means of livelihood for the islanders, Tokyo administration had no other alternative except maintaining the people of Aogashima with a huge subsidy. The inhabitants reared cattle and milk is available in abundance. But that was inadequate to sustain themselves.

I had observed one more surprising spectacle in the island with barely 200 people but having 300 odd cats ! Wherever you stroll on the roads or fields and meadows, there were cats and cats of different colours and sizes. It instantly brought to my mind the horror films of Alfred Hitchkock. When the population of the cats multiplied and outnumbered the human beings in the island, having

The writer meeting Shuchiya san, Governor of Saitama. Ratan Thiyam and Hasegawa san in the background.

nothing to eat, I shudder to think what, if the hungry cats eventually fall upon the inhabitants of the island for their lunch and dinner?. What was particularly baffling to me was the purpose for which the organisers had taken all the trouble to ferry us by helicopter at exhorbitant cost to give just one performance for hardly 200 people there. Hasegawa san explained: cultural activities need not necessarily be the privilege of the city-dwellers alone. The citizens living in the far-flung corners of the state equally deserve their share.

There is still an active volcanoe in the island and we went for a visit there. The area was cup–shaped, surrounded by lofty hills formed of solidified lava. Smoke and steam were emanating from the volcanoe which were indicative that the volcanoe is still alive. That place was being developed as a tourist spot. On the foot of the volcanoe, there was a SAUNA-BATH. The hot steam was chanelled by pipe upto the Sauna Bath. Inside the Bath, a thermometer was fixed to measure the intensity of the heat. One can adjust the temperature of the steam according to one's need and desire. The Sauna Bath is quite popular since the steam emanating from the volcanoe is beleived to possess healing power for gout and other bodily ailments, and refresh the body. When every tourists were doing, why not we too, indulge ourselves in a Sauna Bath, following the age-old adage, 'when you go to Rome, do as the Romans did'.

On October 6, the two teachers who spoke Hindi (Ms. Mitome and Hiuchi Akinari) took us for a guided tour of the island. We could see a rare technique of harvesting rain-water. There was no source for drinking water in the island and as such water was the biggest problem for the population. All the rain-water in a year were collected and stored in huge reservoirs numbering about three or four. The rain water was purified and filtered to make it potable and ready for distribution to the inhabitants. In certain years of scanty rainfall, drinking-water used to be shipped from the near by island of Hachijyojima or the Tokyo City to make it available to the people. It was most interesting and amazing sight to observe the method of harvesting rain-water in Aogashima. I would crave the indulgence of the readers and allow me to describe in a little more detail, the unique water-supply system of the island. On the highest point of the island, there stood a small hillock, which had been cleared of all vegetations and plastered with cement from top to bottom. The size of the hillock would be about one square kilometre. Rain-water falling on this cemented hill, was channelised and collected in Reservoirs at the foot of the hill. Coming from a place where a bag of cement costs anything between Rs. 200-300, we were simply wonder-struck to imagine the mind-boggling cost of the cement used for plastering an entire hill. This indicated that irrespective of the cost involved, the essence of the economic policy of Japan is the welfare of the people.

Next, we were taken to the Power House which supplied electricity to the islanders. The generators were operated by Diesel oil, which, at a 50% subsidised rate, cost 200 yen in the island. This was how electricity was made available to each and every household. Inspite of the fact that the island had a population of a mere 200 people, it had all the amenities of life available in a big city. Cars, Pick-up vans, Trucks etc. were seen plying on the roads. Even bull-dozers for road-construction work were seen there. There were facilities for ISD telephone from this tiny island in the midst of the ocean. Unlike India, you cannot insert a coin for use of Public Telephones in Japan nor do have the system of computer billing of Telephone calls. You have to purchase a telephone card first for an amount depending on your pocket. This card had to be inserted to the telephone-box for every call. One has to keep such a card handy wherever one travels. Such a card could be repeatedly used for 2 or 3 times, depending on the duration of your call. Similar facilities had not been introduced in the cities of India.

The weather is most unpredictable in this island. The atmosphere and temperature would suddendly drop-down and start drizzling. Since the island stood much higher than the sea-level, strong winds howled throughout the night. That one night we spent at Aogashima would forever remain etched in my mind as one of the most memorable experience of my visit to Japan.

Our return journey was also arranged by Helicopter upto Hachijyojima where we halted one night *en route* to Tokyo. The next day i.e. October 7, we left for Tokyo by a ship which happened to be not very big. It started shaking left and right. This was an unfamiliar experience for us and all the artistes felt very giddy and laid down on their beds. They would rather forgo their lunch than get up and eat. This was nothing but sea-sickness which was quite a new experience for all of us. The ship was shaking violently when it approached Miyakojima island, due to the presence of a strong sea-current there. After crossing that area, the shaking stopped and the ship became calm and stable. We arrived at Tokyo Harbour at 9 p.m., and then boarded the waiting bus which took us straight to Hanno City. It was already midnight when we reached the city. Arrangements were made for our stay for the night at the private residence of a friend of Hasegawa-san (his name was Akio Mitshuhasi). We had nothing particular to do at Hanno city; it was only a stop-over enroute to Saitama prefecture which was our real destination. We had an appointment with the Governor of Saitama, Mr. Suchia-san, on October 9. He also happened to be the President of the Indo-Japan Friendship Association. He warmly welcomed the internationally renowned Theatre Director, Ratan Thiyam. Recalling the long friendly relation between the two countries, Ratan Thiyam gave a short speech thanking the Governor for his warm reception of the artistes. We had tea together with the Governor who exchanged pleasantries with the artistes. He further informed us that he would be visiting Imphal on 5 November to pay his homage to the Japanese War Memorial (he actually came to Imphal via New Delhi). He gave small presents to each and every artistes - gold chains for the girls and 50,000 yen for sundry expenses of the Troupe.

It was already 10 AM when we came out after meeting the Governor. Our scheduled programme of performance was a bit upset due to the sudden fixture of a meeting with the Governor of Saitama which was not pre-planned. We had our scheduled programme at Niigata prefecture at 7 p.m. the same night. And on the other hand the meeting with the Governor too, was equally important and unavoidable. It was impossible to cover the distance between Niigata and Saitama by our bus and present a timely performance of our programme at 7 p.m. Cancelling our programme at Niigata was also out of question. In such a predicament, Hasegawn-san provided the solution.

The only way to reach Niigata in time for our performances was to travel by the famed 350 km per hour Sinkansen (better known as the Bullet Train). We have abandoned our bus and alighted the Sinkansen. Upto this point, Ratan Thiyam also accompanied us. The Sinkansen was an incerdibly fast train. We felt as if we were flying in an aircraft. In terms of neatness and cleanness, there is nothing like it. The train was aptly called Bullet Train is because it flew like a bullet. I failed to fathom fully, since I am not a student of science, as to the mechanism of the train which enabled it literally to fly. But as per the information I gathered, I learnt that a magnetic field was created by passing electric current through the iron rails and the train ran (flew!) about two or three inches above the tract without touching it. I was no better informed; the more they explained the more baffling it was to me. We came 'flying' in the Bullet Train and it was hardly 1p.m. when we reached Niigata, well in time for our performance.

Sinkansen (Bullet Train)

At Niigata, we stayed at the Miyokoji Buddhist Temple in Kakudahama City. The priest of the temple (Eiji Ogawa san) and his jean-clad wife received us warmly. That night we were introduced to each other and had dinner together with them.

In fact, a perfect harmony was maintained between god and men in a Buddaist Temple. And what a difference with our temples! Descendants of the royal family, the Chief Minister of the State and other sundry Ministers are permitted to enter the sanctorum of Govindji Temple to which ordinary mortals are debarred from approaching even upto the the temple door. A god who discriminates between differennt categories of people, is not my ideal god. To me, real God is one who cares for the poor and the down-trodden on earth. But god had never dictated that there should be discrimination. Me thinks, it is all artificially created by His 'devotees'.

During our stay at Miyokoji Temple, many Japanese scholars and teachers came to meet us. They were interested to know about the people and culture of Manipur and expressed their desire to visit the state. One of the visitors with whom I interacted was Yoshihiro San, Assistant professor of Music, Niigata University. (He actually came for a visit to Manipur in December, 1995 and stayed there for five days). On October 11, we took leave of Priest Ogawa-san and his gracious wife and left Miyokoji

Arrival of Artistes at Aogashima by Helicopter.

Temple at 3 p.m. for Murakami City. Our dance performance was scheduled for the same night in that city. From Miyokoji Temple, Ratan Thiyam and our Dance Troupe parted company. He had been escorted by Hassegawa-san by car upto Tokyo, wherefrom he would proceed to Hiroshima by train. He would be conducting some research there for his forthcoming play 'Hiroshima' and after spending a few days in that city, he would fly to Imphal. But my group had to stay a few days more in Japan to complete our itinerary. Our next stop was at Murakami city.

30

MURAKAMI- SIRO-SHIBATA

From Miyokoji Temple of Kakudahama upto Murakami was a drive of about two hours only. We reached there at 5 p.m. As usual, after giving our performance, we drove for about 30 minutes and stayed for the night at a place called Siro. It was almost midnight when we finished cooking and had dinner. Making Siro as our Head Quarter, we performed in many neighbouring towns and cities.

The next day (October 12) our party gave performance at a small town of Siro city named Shibata. Their Hall was quite big, having a seating capicity of a thousand people. But on that night the attendance was poor. Most of the audience were elderly people. After the performance, we returned to Siro around 10 pm. The normal duration of our performance was only one and half hours.

It was now more than one month since we left Oike village and Mithila Museum, our temporary home in Japan. After travelling every nook and cranny of Japan continuously for one month giving performances of Manipuri dance and promoting the unique cultrure of Manipur, we were back to Mithila Museum. All other subsequent programmes were conducted making Mithila as our base. We all got up the micro-bus and proceeded towards that city. The drive from Shibata to Mithila took hardly two hours. We reached there at 9 am., took our meal and took rest till 3 p.m. That very day our performance was scheduled at Yamamoto city. We all got up the Micro-Bus and proceeded towards that city.

31

YAMAMOTO

On reaching Yamamoto, we found quite a big gathering of people waiting eagerly to watch the famed Manipur dances. People from nearby towns also came to Yamamoto. The day was worth remembering especially because I met two veteran Japanese soldiers who had taken part in the Battle of Kohima and Imphal during World War II. On learning that a group of artistes from Manipur had come to Yamamoto, they came all the way from a neighbouring city just to meet us. Seventy three years old Yumoto-san narrated his experiences to us. He was a little over 20 when he came as a member of the Kohima column under general Sato and took part in the fierce Battle of Kohima. He stayed there for about 2 months and came to Imphal. The name of the other elderly soldier was Takejȯ Takahashi-san. He too, was also a part of the Kohima Column and took part in the fighting there. He told us that twice he came upto Imphal.

The Battle of Kohima, lying midway on the Imphal-Dimapur. Road, was one of the fiercest and bloodiest one in World War II. Japanese 15th Army of Lt. Gen. Mutaguchi Renya, G.O.C. had three Divisions viz. 15, 31 and 33 under his command. He issued orders to Lt. Gen.

Sato Kotuku, commanding 31 Division to advance directly to Kohima from his base at Homalin, Burma, to capture it and prevent reinforcement from moving south to relieve the Imphal Garrison of the Allied Force. Gen Sato, accordingly moved his Division from its base to Ukhrul and from there proceeded northward to Kohima through thick jungles and deep ravines where there was no road at all. It was near impossible task, testing human endurance and courage upto the extreme. Food and ammunition were problem, but the Japanese soldiers were used to living hard. The lack of road through which the 31 Division had to pass, being a jungle country, would make it impossible for supplies to be sent up in motor transport. Hence their soldiers carried a month's supply with them during their march to Kohima.

The task assigned to Gen. Sato and his men was to cut-off the Imphal-Dimapur road and thus prevent all supply to reach the Allied HQ. at Imphal. A sanguinary battle was fought at a Tennis Court around Kohima in which the Japanese soldiers were not only defeated but completely routed on 8th April, 1944. The disaster at Imphal (including Kohima) was perhaps the worst of its kind yet chronicled in the annals of war. The ranks of the retreating column had thinned down daily as thirst and hunger overtook and the wounded had to be abandoned by hundreds. The suffering of the Japanese soldiers were beyond description. There was no food except what they could pick up in the Naga villages; no ammunition came up from the rear and there was no medical attention for the sick and wounded. In such a terrible condition, very few could reach their destination. Their defeat was complete and irreversible.

How exciting it was, to meet Yumoto san and Takejo Takahashi san, who happened to be among the handful of survivors of the retreating column of the 31 Division of Lt. Gen. Sato who fought in the Battle of Kohima in April 1944.

After our performance the organiser did us a great honour by giving a reception and dinner party. It was already 11 pm. when we returned to our 'Home'-Mithila Museum.

32

NITSU

The next day (October 14), we left Oike village for a small village/town called Nitsu. Yoshihiro Arai, the driver of our Micro-Bus and our constant companion during our 3 month's tour of Japan, belonged to this village. His parents lived there. He was particularly close to our artistes, having shared all the pleasures and pains of the journey by Bus. Everyone was thrilled and excited at the news that we would be going to the village of Yoshihiro-san. Nitsu was, indeed, a very tiny village; there was no suitable place for our performance and therefore a performing space was improvised inside a gymnasium. We presented our programme at 2 p.m. All the audience who came to watch our cultural show sat on the wooden floor of the gym cross-legged, for no chairs had been provided for them. Considering the size of the population of Nitsu, the attendance was quite encouraging, infact, beyond our expectations. After the performance was over, we spent the night at Nitsu. And incidentally that night at Nitsu turned out to be the most significant and memorable one. Being a small town there was no suitable place like a Hotel where all our nineteen artistes could be accommodated at one place. Accommodation in Japan was always a problem.

Arrangement for our stay for one night was therefore made in a different way. The 19 artistes were divided into small groups of three or four and stayed with different families of Nitsu. Having spent three months in Japan, by now we had picked up the basic Japanese words, enough to communicate with our hosts.

Every one of us were fully conversant with the customs and manners of the Japanese people. We had learnt how to use chopsticks for eating our meals, to maintain the standard of keeping neat and clean, the 'rituals' of going to toilet and kitchen etc. Hence, we were quite at home staying with a Japanese family. At least for our seven female artistes, arrangement were made for them to stay together in one place. But the rest were distributed to five or six families. Myself with another artiste, stayed with Hiro-san (who drove our Bus) and his parents. I shall forever cherish the memory of the loving care and regards his parents shown us, as if we were high dignitaries. That was the greatness of the Japanese.

On the other hand, it was never a problem in Japanese houses to accommodate three or four guests. This was because they did not use charpoy (cot) for sleeping - every one slept on the plank-floor by spreading bedding, blankets etc. In the morning their beddings were rolled up and put inside the built-in cupboards to make the space available for use as drawing room during day time. For us, accommodating four or five

Bidding farewell to the parents of Hiro san.

guests at a time always posed a big problem because, according to our custom and traditions, each of them had to be provided with a cot, well-equipped with bed-rolls, blankets, mosquito nets, bed-sheet and pillow.

That night I skipped dinner and had very little sleep, too. I had received the sad news of the untimely demise of my younger brother Joykumar whose shraddha ceremony (obligatory religious ritual of the Hindus practised on the 12th day of death of a person) happened to be the next day. It was impossible for me to take part in the Shraddha Ceremony, being in far-off Japan. Since I could not forsake my responsibilities as Troupe Leader, I tried to console myself. I did not share this sad information with the rest of the artistes lest it would unnecessarily upset them. The next day, we bid good-bye to the parents of Hiro san and left Nitsu at 8 a.m. for Meabash City of Gumma Prefecture.

33

MEABASH

As soon as we arrived at Meabash at 11 a.m. we were rushed straight to an Indian Restaurant, its owner (J. Shingai san) was a friend of Hasegawa san. He invited us for lunch at his Restaurant. The food served on that day were our favourite dishes. All the cooks of the Restaurant were also Indians, too. We were served with rice and fish while some among us took Nan-Ki-Roti.

Among the people who came to witness our programme, I met an American artiste, Mr. T.A. Hoffman. He married a Japanese girl and had been doing research on Japanese music. He was one of those few artistes who had contributed towards awareness of Hindusthani Classical Music and its promotion in Japan. Mr. Hoffman learnt Hindusthani Classical Music for 5 years in Bhatkhande Music College of Lucknow, India. He told me that he knew a Manipuri: Laishram Mema, who was with them in the Bhatkhande Music College.

That night after our performance was over, arrangement for our dinner was made in a Restaurant called "NEW DELHI". This Restaurant, too, belonged to Hasegawa-san's friend, J. Shingai-san. He had two such Indian Restaurants in this city. The dinner was well laid out with chicken curry, Mutton curry, Kababs ... ... etc. I shared a table with Mr. Hoffman and his wife which provided an opportunity to resume our discussion. I was given to understand that he would be going to Pune, India, next year and expressed his desire to visit Manipur also, (but he could not make it). After dinner, we got on our Bus and left Meabash. It was already past midnight when we reached our 'Home' Mithila Museum.

We took rest the next day. We had by then nearly completed our itinerary in Japan. There was only one show left - at Sanjo City.

Tim Hoffman playing the Japanese wind-instrument Saku hachi.

34

SANJO

On October 17, at 3 pm. we started for Sanjo. It happened to be a small village. The organiser of our programme in this village was a lady, Mrs. Kawase-san. Her husband was a doctor who had opened his own private clinic. Quite unrelated with their professions, both wife and husband were connoisseurs of art and culture like dance and music. The couple organised our cultural show by selling tickets.

When our performance was over, Mrs. Kawase-san arranged a reception for the artistes in the clinic of her husband. After the reception, we presented her a small souvenir to remember Manipur- a Khamba Thoibi doll. The hospitality of the Japanese were amazing; wherever we went the people literally make a bee-line to give us small presents. They never asked for anything from us. But alas! when Japanese tourists visited Manipur many unhappy and unseemly incidents were reported to have happened. It was rumoured that their cameras, cassette-recorders and even their hats were stolen. We are, by and large, averse to giving presents to others; we literally solicit presents from others. We returned the same night to Mithila Museum.

35

MITHILA MUSEUM (3)

In Japan, we had given more than forty performances within a span three months. Over and above these normal programmes, we had conducted a number of workshops of Manipuri Dance and Thang Ta (Manipuri martial arts) in many cities which were unaccounted.

On the penultimate day of our stay in Japan, we went shopping at the nearby city Tokamachi. On our last night at Mithila, Hasegawa-san arranged an elaborate farewell-party for the Manipuri artistes inside his Museum. He invited some of his friends also in that party. The two terracotta artistes from Bihar, Lala Pandit and his brother, Jagdish, the two Madhubani painting sisters, Smt, Godavari Devi and Shashikala Devi, were also present at our farewell-party. Many renowned artistes and artisans were invited from time to time to Mithila Museum and

Farewell Party at Mithila Museum. Terracota figures in the background.

their creative-works were displayed there. The well known pottery-artiste, Smt Nilamani Devi from Manipur, was also invited by Hasegawa san in 1998 for making traditional Manipuri potteries for his Museum. Till the day, when we finally left Japan on October 29, 1995, the two terracotta artiste-brothers continued their work in Mithila Museum.

It was customary in Japan to raise a toast at the beginning of such parties. Each one of the people present there would raise a glass of beer (or a soft drink) in their hands. The leader or the host would utter "Kampai" when everyone responded 'Kampai' in unison and started the drinking session. It was also customary to make short speeches from the host and the guests. I spoke briefly and lauded the works of Hasegawa-san, particularly his initiative in popularising Manipuri dance and culture in Japan. I further stated that for his role in exposing the culture of Manipur, not only in big towns, cities and villages but also in remote islands in the ocean, too, I should consider him as the unofficial cultural Ambassador of India in Japan. This was not an over-statement as an act of courtesy. Many cultural troupes from Manipur had earlier visited Japan and performed there. But the difference was that those earlier groups came to Tokyo by air, gave one or two performances in some big cities, stayed in posh Hotels for a few days and returned home.

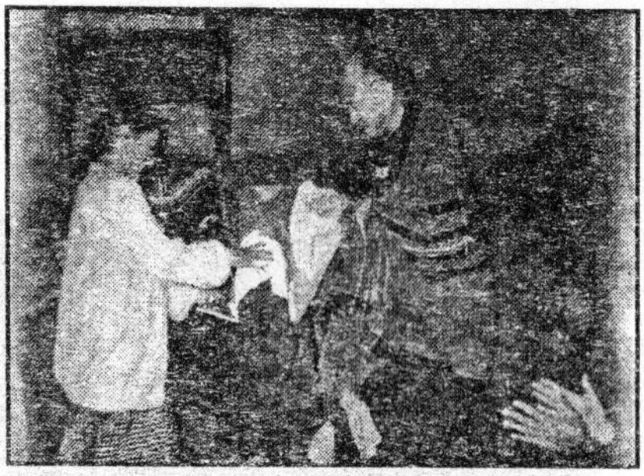

A farewell present to Hasagawa-San

They never had the opportunity to travel the length and breadth of Japan and closely interact with its people. Our cultural troupe organised by Ratan Thiyam was the first group which had given performance of Manipuri dance in every nook and cranny of Japan. During these prolonged and hectic tour, we hardly had any rest. At times we endured many physical hardships. The tiring long journey of thousands of miles by bus, our whole bodies aching, no proper accomodation or fixed time for our meals - these were some of the discomforts we encountered. For all these small inconveniences, however, we had been amply rewarded in another way. The sound of Dhol Dholak and the strains of Manipuri folk music had resounded the Pacific Ocean and the Sea of Japan.

On the night of the farewell party, we presented Manipuri Handloom clothes to our host and hostess, Phanek and Rani-Phi for Mrs. Hasegawa. Our girls helped her to dress up in this Manipuri attire. She looked simply stunning when she appeared at the farewell party. Hasegawa san presented me, two Samurai Swords with love which now occupy a pride of place in my drawing room.

"Kampai"

36

SAYONARA JAPAN

On October 19, after having an early brunch, we boarded our micro-bus at 11 a.m. and headed for the Narita International Airport of Tokyo. At 7.30 p.m. we left Narita Airport by a Singapore Airlines flight. The two Japanese youths, Hiro san and Kiuseka-san, who had been our constant companion during the past three months broke down at the moment of our departure. Perhaps, this would be the last time we would see them again in our life. All our artistes also wept and bade them a tearful farewell.

Our aircraft had a non-stop flight throughout the night and we arrived at Changi Airport of Singapore early the next morning at 11 am. We had to stay the whole day at the airport since our connecting flight to Calcutta would leave only at night. Everyone felt hungry. Arrangement of meals for 19 of us was made at a Restaurant inside the Airport. It was a simple frugal meal consisting of a plate of rice and fish, but was so expensive - the Bill came to about Rupees six thousand. It was not an elaborate meal; yet each one had to pay about Rupees 359/- which, by Manipuri standard, was quite high.

Changi Airport is reputed as one of the best and well-equipped airport in Southeast-Asia. It was difficult to walk down the whole length of the lounge from one end to the other due to its sheer size. Passengers had to use escalators for reaching from one corner to another. There were a number of duty-free shops inside the airport. A wide range of wares were on sale; in fact, nothing was lacking which a shopper needed. One can buy these goods tax-free by producing one's passport. Foreign tourists, especially the Americans, made big purchases. Those shops were, in fact meant for big-budget tourists with their pockets full of dollars, not for people like us who came from a poor state like Manipur. By our standards, the articles were priced exorbitantly high and beyond our reach. Besides, there was nothing which could be purchased by Indian currency (Rupee). Only those

Royal family of Japan (From Left) Prince Naruhito, Emperor Akihito, Queen Michiko, Princess Sayako and Prince Fumihiko

persons with hard currency could afford it. We had to stay the whole day at Changi Airport. Awaiting our night flight and taking advantage of our forced rest at the Airport for about four to five hours, I went for sight seeing, along with a select few artistes, to the Singapore City. I thought it prudent not to take the entire group of 19 artistes lest it would be unmanageable and would not be able to return to the Airport in time for our scheduled flight. The sight-seeing was a conducted tour by the Airport Authority and no extra charges need be paid. We had to surrender our pass-ports to the Airport for the sight-seeing trip. The tourists were taken in their coach. Singapore city is full of high rise buildings and no natural landscape was in sight. There was hardly enough time for an elaborate sight-seeing programme. Within a stipulated time, the coach took us to the city and dropped in the Airport timely. On my return, the artistes wondered where had I disappeared for four or five hours. I concealed the facts sensing that they might take offence for being left out of the sight-seeing programme. Today, through the pages of this book, the mystery of my disappearance from Changi Airport with a handful of boys is being unravelled. The same night, we flew from Changi Airport to Calcutta at 10 p.m. and halted in that city for two days before we returned to Imphal on the 22nd October, 1995.

37

EPILOGUE

Since we had lived with the Japanese people for a pretty long time, we had a smattering of their language and could speak it too. We also had a fairly good knowledge of their habits and customs. The Japanese are polite and polished in their behaviour. In Japan, the first lesson of a child was to utter ARIGATO (meaning Thank you in English). But in Manipur, being the land of the Gandharvas, (celestial musicians) the first lesson of a child is dancing. I recall an incident while in Japan. Some of us went out for a stroll in the morning. A young boy of 7 or 8 years was playing with his bicycle. Suddenly he fell down from his bike and was hurt. The boy was crying in the street. One of our boys rushed and helped the child to get-up. In between sobs, the boy spelled out ARIGATO. The incident, though a trifle and insignificant, speak volumes about the Japanese traits of humility and deceney.

Yet another trivial incident comes to my mind. We were walking in a street. Suddenly a telephone number popped up in my mind and I needed a piece of paper to jot it down. All of us searched for a piece of paper throughout the stretch of the street. Believe it or not, we could not find even a smallest piece of paper littering the street. It was quite a surprise to me. For us who came from India, where the streets were littered with rubbish, it was short of a miracle to come to a land where the streets are maintained so tidy without even a piece of paper on them. The most interesting aspect is that no sweepers were also employed to sweep the streets, as we do in India. Then how was it possible to maintain the streets spick and span? The secret lies in the voluntary action of the people. In Parking lots, Parks and others public places, Tourist Centres and Restaurants people were seen holding cellophane bags in their hands and religiously picking up shreds of paper or other rubbish materials and put them inside their bags. The dirts, thus collected by the citizens, were dumped into dustbins. Everyone seemed to be proud of doing so. Since every citizen felt it their civic duty to keep their cities clean, little wonder that Japan is the most neat and clean country in the world.

FOOD : Japanese food are quite different from ours. We are never comfortable and satisfied with the type of food they eat. We, the Meiteis, eat literally to fill the stomach; a nation averse to eat only a small quantity of high protein and calory filled food for the sake of health. Even after eating almost 9 to 10 chapatties, we would again insist on rice (Rice is the staple food of the people of Manipur). Their reasoning is: chapati (baked wheat cake) is chapati, not rice. No Manipuri meal is complete unless supplemented by a liberal amount of rice. To us only rice is real food; the rest just appetizers!

The basic difference between the Japanese food and ours is in the method of cooking; the ingredients are the same in both the countries. Like us, their staple food is also rice and fish. The Japanese rice contains a very high percentage of starch. When a guest is invited for lunch or dinner, they would invariably serve rice and fish. But the most widely used food item and in variegated forms is Soyabean. There are a number of foods prepared from Soyabean. During a meal, they would serve Soya soup (Miso), Pickles, a few pieces of fish / meat, Seaweeds called Nori, somewhat akin to our Sugnoo Nungsham (water algae). They toast this sea -weed over fire and serve to the guests. That consists of a full Japanese dish. For Japanese, meat is not their traditional food; it is fish. But nowadays beef and pork are being liberally used in Japanese homes.

There is yet another way of serving fiood. Rice is made into balls over which thinly sliced raw fish (they call it Sashimi) are placed and Soya soup poured over it. I hardly come across any Japanese drinking water during meals. On the contrary, they liberally drink beer. After food, they love to take green tea (which they call Ocha). Their tea is without milk or sugar. Our tea with milk and sugar is known as English tea in Japan. So be careful while in Japan; if you ask for a cup of tea they would serve you pure tea-liquor. The most popular and renowned drink in Japan is Saki, prepared from rice. It is just like our rice beer and is very mild, slightly sweet and has very little alcohol content. Generally it is sold in a big bottle almost double of our standard wine bottle.

Another food item which is widely used with relish is Hawai-Achar (they call it Natto). It is sold in sealed packets in every shopping centre and Departmental stores. Especially in Tokyo city, it is much in demand. Method of preparing Natto in Japan is quite different from our method. We boil the soya bean and then keep

Sashimi

it for a few days in an airtight basket, for natural fermentation. But the Japanese apply yeast powder to the boiled soyabean which instantly ferments it. The taste seemed different from our own brand, probably because of the instant fermentation. But the Japanese method may be better for health due to minimum loss of food value.

KABUKI : Kabuki owes much to both 'No' and 'Kyogen', the main theatrical forms of the medieval age. The acknowledged "originator" of Kabuki was a woman named OKUNI. Okuni led a troupe of female dancers in Kyoto in a kind of outdoor musical entertainments that was labelled (by others) "Kabuki dancing". The term Kabuki was derived from KATAMUKI"- slanted or strangely inclined. In 1629 AD. Shogunate banned the participation of women in Kabuki altogether. This gave rise to young man's Kabuki. This aroused homo-sexual passions. Therefore the Shogunate also banned these handsome young men to perform Kabuki.

Henceforth, only adult males were allowed to perform on the Kabuki stage. The banning of women from Kabuki gave rise to the unique personage of the ONNAGATA, or male performers of female roles. So special are the acting qualities cultivated over the centuries by the Onnagatas that this Jápanese art form has become unique in the world.

In Manipur, too, there is a strong tradition of female roles being played by males in our Shumang Lila (courtyard plays). And truth to tell, the Nupishabi (males in female roles) of our Shumang Lilas are far superior to their counterpart in the Japanese Kabuki Theaters in term of beauty and artistic excellence.

BUNRAKU (Puppet Theatre) : The mature art of Bunraku, as it was developed in the late 16th and 17th century, has been defined by Donald Keene as "a form of story - telling, recited to a musical accompaniment and embodied by puppets on a stage". Of the three main elements of Bunraku - story-telling, musical accompaniment, and the use of puppets - it is the story telling that is of greatest importance in the history of Japanese culture. Puppetry was a minor theatrical form that was used to supplement the traditionally derived art of the Bunraku chanters. Chanters generally used a mandolin-like four-stringed instrument called BIWA. Subsequently, a three-striged instrument called SAMISEN (originally from China) came into vogue among chanters.

The Manipuris also have a long tradition of story-telling, although the method is slightly different from that of Japan. There are professional story-tellers who perform in temples during religious occasions and they render the heroic tales from our ancient epics like the Ramayana and the Mahabharata.

HAIKU: Basho adopted as his principal medium of expression, the Seventeen syllable HAIKU. Certainly, one of the world's briefest verse forms, HAIKU derives from the first phrase or link of the classic TANKA and consists of three lines of 5-7 and 5 syllables. Since the rules are simple, almost anyone can compose these 17- syllable poems, and indeed Japanese of all classes have written HAIKU through the centuries from

Basho's time. But the HAIKU is something like the ultimate in deceptive simplicity, and out of a vast number of "acceptable" ones only a fraction are apt to be truly fine.

Basho's Haiku is of such an extra-ordinary quality as to make him without question one of the greatest Japanese poets. With little more than a handful of syllables at his disposal, the writer of Haiku obviously cannot hope to enter into extended poetic dialogue . He must seek to create an effect, to capture a mood, or to bring about a sudden and sharp insight into the truth of human existence. His best known Haiku, for example is

"An ancient pond
A frog jumps in
The sound of water"

Basho had not said how wondrous it is to observe the meeting of that which is eternal, as embodied in the ancient pond, and that which is fleeting, as represented by the frog's jump. In the best Zen and Haiku spirit, he has simply juxtaposed the two images without subjective comment and has left it to the reader to draw whatever meaning or meanings he can from the poem.

Some more Haikus from Basho's pen

(ii)

"The summer grasses
Of brave soldier's dreams
The aftermath,"
(From the Narrow Road of Oku)

(iii)

"Such stillness -
The cries of the Cicadas
Sink into the rocks".

38

RELIGION OF JAPAN

Buddhism came to Japan only by the end of sixth Century A.D. via China. Before that for thousands of years, the Japanese followed their indigenous religions called Shintoism. In the present time, the Japanese follow a synthesis of the two religions.

In Japan, the native mythological assertion is that the Sun Goddess has granted a mandate to the imperial family to rule eternally. According to mythology, the ruling dynasty of Japan is descended from the Sun Goddess (AMATERASU), the supreme deity or KAMI of the Shinto pantheon. The Sun Goddess gave dominion over the plain of high heaven, She directed her grandson, NINIGI, to descend from heaven and establish the rule of his family on earth. At the time of Ninigi's departure, the Sun Goddess bestowed upon him the sacred regalia mirror, sword and jewel as token of his mandate

Marraige in Japan according to Buddhist rites. The bride is drinking 'Sake'.

to rule. In Japanese tradition, it is solely on the basis of this mandate that members of the imperial family have legitimately reigned from the founding of the Empire in 160 B.C. We know very little about the evolution of those religious beliefs of ancient Japan that collectively came to be called Shinto (the way of the Kami or gods). In Shinto we can observe a primitive religion of the sort that elswhere in the world has been absorbed by the universal faiths but that in remote and parochial Japan, it has been perpetuated into modern times.

What is Shintoism? The central feature of Shinto is its animistic belief in the Kami spirits that fill the world, inhabiting living things as well as mountains, rocks, streams and so forth. The Kami have always been essentially the object of local community and family devotion; indeed, their special character is the intimate association they have with these fundamental entities of Japanese social organisation. The very word Kami has the connotation of 'upper' or 'above'. I am tempted to draw a parallel with Atiya-Guru Shidaba of the Sanamahi cult of the Meities. Another point of similality between our Sanamahi cult and Shintoism is the sprinkling of sacred water before entering the Shrine. Purification is particularly important in Shinto; in fact, it is the principal act performed at Shinto Shrines; usually by the sprinkling of water with a wand. We the Meities, also practice such a similar ritual while entering the sacred shrine of our traditional deity known as Umang Lai by sprinkling sacred water with the leaves of Tairen (a local tree; its scientific name , *Toona ciliata*). The most familiar structure of Shinto common to all shrines is the entrance way or TORII, literally meaning 'bird perch'.

All the Japanese who follow Buddhism today do not abandon their age-old Shinto religion; they still follow the rites and rituals laid down by Shintoism. Herein lies a striking similarity between the religion of Japan and that of Manipur. Every Manipuri, although converted to Goudiya Vainavism, never completely cuts-off the rituals and religious pratices associated with the Sanamahi religion. There is no Manipuri house-hold which does not preserve the Sanctum Sanctorum of Sanamahi and Leimaren. Not only the Meiteis, even the Pangals (Manipuri Muslims) and the Brahmins do the same.

On the sixth day when a child is born, the Manipuris perform Swasthi Puja at night while the ancient Meitei ritual Epan Thaba is also performed in the morning of the same day. Is there a similar tradition in any Hindu sect of India except the Manipuris? Which denomination of Hindu religion follow the ritual of releasing three Ngamus (a type of local fish) in a nearby pond by three women, holding a hurricane lamp in their hands, during a marriage ceremony, except the Meiteis? This is clearly a non-Hindu ritual being adopted by the Meiteis from their ancient Sanamahi cult. This peculiar custom is being followed only by the Manipuri Hindus and never by any denomination of Hinduism in the rest of the country.

But the amazing fact is that this primitive religion called Shintoism is still practised in Japan as a living religion. It is impossible to explain how primitive Shintoism managed to survive the influx of Buddhism. Part of the answer lies in the unusual tolerance of Eastern religious thought in general and its capacity to synthesize seemingly disparate beliefs and manifestations of the Divine. In Japan for example, the principal KAMI of Shinto came to be regarded as Budhist deities in different forms, and Shinto shrines were amalgamated with Budhist temples. Another reason why the Japanese throughout the ages have with little or no difficulty and considered themselves to be both Shintoists and Buddhists is that the doctrine of the two religions complement each other so neatly. It is fitting that even today in Japan the rituals performed to celebrate such events as birth and marriage are Shinto, whereas funerals and communion with the dead are within the purview of Buddhism.

Most Japanese, although devout Buddists, never completely abandon the tenets of Shintoism; both the religions are equally revered. In this respect, we can do well to emulate Japan. If we could treat with equal love and reverence to both Sanamahi and Goudiya Vaishnavism it would go a long way in bringing religious and social harmony in Manipur.

Although Japan today, is considered as one of the most advanced country in science and technology, they have never lost sight of their ancient culture and tradition. The old and the new are being harmoniously blended and well-balanced. Herein lies the uniqueness of the Japanese religion.

PLACES IN JAPAN WHERE WE PERFORMED MANIPURI DANCE

1. 24/7 Akadomari-mura/Niigata/Akadomari-mura.
2. 27/7 Sendai City/Miyagi Pre./ Sendai City.
3. 38/7 ibd.
4. 31/7 Yokohama City/Kanagawa Pre./Executive Committee.
5. 2/8 Bato-cho/Tochigi Pre./International-Exchange Comm.
6. 3/8 Narugo-machi/Iwate Pre./Executive committee.
7. 4/8 Iwaizuma-machi/Iwate Pre./Executive committee & then back to Mithila Museum.
8. 14/8 Matsuyama-machi/Miyagi Pre./Matsuyama-machi.
9. 18/8 Tokamachi City/Niigata Pre./Executive Committee.
10. 20/8 ibd./Akakura Elementary School.
11. 22/8 Tajimi & Halt till 23/8.
12. 25/8 Imabari city/Ehime Pre./Executive Committee.
13. 27/8 Oguni-machi/Kumamoto Pre./Oguni-machi.
14. 29/8 Fukuoka-City/Fukuoka Pre./Fukuoka City.

15. 30/8 ibd. Fukuoka
16. 1/9 ibd. Fukuoka
17. 3/9 Yamaguchi City/Yamaguchi/Youth International Exch. Comm.
18. 6/9 Kagoshima
19. 7/9 Miyakonojyo City/Miyazaki/Executive Committee.
20. 10/9 Atami City/Shizuoka/Executive Committee.
21. 11/9 Yokaichi City/Shiga/Executive Committee.
22. 12/9 Tajimi City/dify/Executive Committee.
23. 15/9 Matsue City/Shimane/Executive.
24. 16/9 ibd.
25. 18/9 Fukuoka city/Fukuoka/Fukuoka City.
26. 19/9 ibd.
27. 20/9 ibd.
28. 21/9 ibd.
29. 22/9 ibd.
30. 23/9 ibd.
31. 24/9 ibd.
32. 28/9 Yokohama City/Kanagawa.
33. 30/9 Yokohama City/kanagawa.
34. 1/10 Sumida-ku/Tokyo/Asia Club.
35. 3/10 Hachijyo-Jima/Tokyo/(expected)
36. 5/10 Aogashima-mura/Tokyo/Aogashima-mura.
37. 9/10 Niigata City/Niigata/Executive Committee.
38. 11/10 Murakami City/Niigata/Executive Committee.
39. 12/10 Shibata City/Niigata/Executive Committee.

40. 13/10 Yamato-match/Niigata/Executive Committee.

41. 14/10 Niitsu City/Niigata/Executive Committee.

42. 15/10 Meabash City/Guuma Prefacture.

NAMES OF ARTISTES WHO VISITED JAPAN

1. RATAN THIYAM 47才
ラタン・ティヤム 団長・俳優
2. CHONGTHAM SAMARENDRA SINGH 58
チョンタム・サマレンドゥラ・シン マネージャー
3. HAOBAM IBOCHAOBA SINGH 才 50才
ハオバム・イボチャオバ・シン ドラマー
4. THOUNAOJAM MEGHACHANDRA SINGH 21才
トウナオジャム・メガチャンドラ・シン ドラマー&ダンサー
5. HAORONGBAM RAJMANI SI NGH 23才
ハオロンバム・ラジマニ・シン ドラマー&ダンサー
6. GURUMAYUM LOKEN SHARMA 26才
ギルマユム・ロケン・シャルマ ドラマー&ダンサー
7. AMON DEVEN SINGH 23才
アモム・デヴェン・シン ドラマー&ダンサー
8. WAHENGBAM BROJENDRO MEETEI 25才
ワヘンバム・ブロジェンドロ・シン 武術アーティスト
9. WAHENGBAM SUNIL SINGH 19才
ワヘンバム・スニル・シン 武術アーティスト
10. WAHENGBAM LOKENDRO SINGH 24才
ワヘンバム・ロケンドロ・シン 武術アーティスト
11. GOTIMAYUM MANISANA SHARMA 18才
ゴティマユム・マニサナ・シャルマ ドラマー

12. NAMEIRAKPAM BOYNAO SINGH 12才
ナメイラクパム・ボイナオ・シン ドラマー&ダンサー
13. YAMBEM SUNILKUMAR SINGH 26才
ヤムベム・スニルクマル・シン フルート
14. THOKCHOM UMARANI DEVI 26才
トクチョム・ウマラニ・デーヴィー ダンサー
15. HUIREM SUSHMA DEVI 19才
フイレム・スシュマ・デーヴィー ダンサー
16. NONGTHOMBAM BABITA DEVI 26才
ノントムバム・バビタ・デーヴィー ダンサー
17. LAISHRAM MANDAKINI DEVI 23才
ライシュラム・マンダキニ・デーヴィー ダンサー
18. HAOBAM SUNITABALA DEVI 18才
ハオバム・スニティバラ・デーヴィー ダンサー
19. NINGOMBAM USHARANI DEVI 17才
ニンゴムバム・ウシャラニ・デーヴィー ダンサー
20. YAIKHOM SAYALAKSHMI DEVI 22才
ヤイコム・サヤラクシュミー・デーヴィー ダンサー

INDEX

■■■